Clemens Schmetterer

Phase Equilibria in Ni-P-Sn and Binary Subsystems

Clemens Schmetterer

Phase Equilibria in Ni-P-Sn and Binary Subsystems

Understanding the Reactions between Sn-based Solders and ENIG Surfaces

Südwestdeutscher Verlag für Hochschulschriften

Impressum/Imprint (nur für Deutschland/ only for Germany)
Bibliografische Information der Deutschen Nationalbibliothek: Die Deutsche Nationalbibliothek verzeichnet diese Publikation in der Deutschen Nationalbibliografie; detaillierte bibliografische Daten sind im Internet über http://dnb.d-nb.de abrufbar.

Alle in diesem Buch genannten Marken und Produktnamen unterliegen warenzeichen-, marken- oder patentrechtlichem Schutz bzw. sind Warenzeichen oder eingetragene Warenzeichen der jeweiligen Inhaber. Die Wiedergabe von Marken, Produktnamen, Gebrauchsnamen, Handelsnamen, Warenbezeichnungen u.s.w. in diesem Werk berechtigt auch ohne besondere Kennzeichnung nicht zu der Annahme, dass solche Namen im Sinne der Warenzeichen- und Markenschutzgesetzgebung als frei zu betrachten wären und daher von jedermann benutzt werden dürften.

Verlag: Südwestdeutscher Verlag für Hochschulschriften Aktiengesellschaft & Co. KG
Dudweiler Landstr. 99, 66123 Saarbrücken, Deutschland
Telefon +49 681 37 20 271-1, Telefax +49 681 37 20 271-0, Email: info@svh-verlag.de
Zugl.: Vienna, University of Vienna, PhD Thesis, 2009

Herstellung in Deutschland:
Schaltungsdienst Lange o.H.G., Zehrensdorfer Str. 11, D-12277 Berlin
Books on Demand GmbH, Gutenbergring 53, D-22848 Norderstedt
Reha GmbH, Dudweiler Landstr. 99, D- 66123 Saarbrücken
ISBN: 978-3-8381-0997-8

Imprint (only for USA, GB)
Bibliographic information published by the Deutsche Nationalbibliothek: The Deutsche Nationalbibliothek lists this publication in the Deutsche Nationalbibliografie; detailed bibliographic data are available in the Internet at http://dnb.d-nb.de.
Any brand names and product names mentioned in this book are subject to trademark, brand or patent protection and are trademarks or registered trademarks of their respective holders. The use of brand names, product names, common names, trade names, product descriptions etc. even without
a particular marking in this works is in no way to be construed to mean that such names may be regarded as unrestricted in respect of trademark and brand protection legislation and could thus be used by anyone.

Publisher:
Südwestdeutscher Verlag für Hochschulschriften Aktiengesellschaft & Co. KG
Dudweiler Landstr. 99, 66123 Saarbrücken, Germany
Phone +49 681 37 20 271-1, Fax +49 681 37 20 271-0, Email: info@svh-verlag.de

Copyright © 2008 Südwestdeutscher Verlag für Hochschulschriften Aktiengesellschaft & Co. KG and licensors
All rights reserved. Saarbrücken 2008

Produced in USA and UK by:
Lightning Source Inc., 1246 Heil Quaker Blvd., La Vergne, TN 37086, USA
Lightning Source UK Ltd., Chapter House, Pitfield, Kiln Farm, Milton Keynes, MK11 3LW, GB
BookSurge, 7290 B. Investment Drive, North Charleston, SC 29418, USA
ISBN: 978-3-8381-0997-8

"Does anyone have information about Ni-P-Sn?"

COST 531 WG1+2 meeting in Porto, 25th October 2004. That was the day all of it started…

"You will never get anything useful from Ni-P-Sn."

Comment just before the first SEM measurement in the ternary system. Well, at least we got some data – if they are useful the future will show (probably).

"Calorimetry in Ni-P could give some entirely new meaning to the word Calorimeter Bomb."

One of our thoughts we gave to seriously starting calorimetric measurements in Ni-P.

Acknowledgements

This work was accomplished at the Department of Inorganic Chemistry / Materials Chemistry of the University of Vienna from September 2005 to March 2009.

In the first place I would like to thank my parents, my grand parents and my brother for all their interest, their support and their encouragement during my studies.

I want to express my greatest thanks to Prof. H. Ipser for the opportunity to write my PhD thesis in his institute, for all his support, his readiness for discussions and his help during the turbulent closure of the ELFNET project that once more proved that complex scientific issues are still more logical than finance maths.

Special thanks go to Prof. Hans Flandorfer and Prof. Klaus Richter for their readiness to help with all theoretical and practical difficulties and for being open for discussions. I also want to thank Dr. Klaus Richter for getting my exuberant crystallographic ideas back on track and for his help with crystallographic matters in general.

I am grateful to Prof. Gerald Giester and Prof. Manfred Wildner for performing the single crytstal X-ray diffraction experiments and for their assistance with the structure solution.

Thanks to all colleagues and collaborators in the Institute of Inorganic Chemistry – Materials Chemistry for the pleasant working conditions, especially Mag. Norbert Ponweiser who had to endure extra shifts with me at the EPMA to get my Mt. Everest-like amount of samples measured, Dr. Jiri Vizdal who set up sample preparation without blowing up the furnace, and Dr. Rajesh Ganesan who was not afraid to join the phosphorus-team with the isopiestic method.

I also want to thank Prof. Alexandre Kodentsov from Technical University Eindhoven, who hosted me for one month during a Short Term Scientific Mission within COST 531 and helped me with the first set of SEM investigations.
I am thankful to Dr. Adela Zemanova and Dr. Ales Kroupa from the Academy of Sciences of the Czech Republic for carrying out additional SEM measurements.

The financial support of the Austrian Science Foundation 'Fonds zur Förderung der wissenschaftlichen Forschung' under Project No. P18968-N19 is gratefully acknowledged. This thesis also contributes to the European COST Action MP0602 HISOLD and has already contributed to COST Action 531 and ELFNET. Further thanks go to ÖAD, Austrian Academic Exchange Service for granting cooperation with the Institute of Physics of the Czech Academy of Sciences within bilateral Project 07/2007.

Finally I would like to thank all my friends for all their understanding and for making my life enjoyable.

Table of Content

1. Introduction 1
   1.1 Principle of Soldering 1
   1.2 The Influence of Process Parameters with the Focus on Lead-Free Soldering 2
   1.3 Contemporary Soldering Techniques 4
   1.4 Phase Diagrams and (Lead-Free) Soldering 5
   1.5 The Role of the Ni-P-Sn Phase Diagram 6

2. Literature Review 11
   2.1 The Binary System Ni-Sn 11
   2.2 The Binary System Ni-P 13
   2.3 The Binary System P-Sn 15
   2.4. The Ternary System Ni-P-Sn 15

3. Experimental Section 18
   3.1 Preparation of Binary Ni-P Alloys 18
   3.2 Preparation of Binary P-Sn Alloys 19
   3.3 Preparation of Ternary Ni-P-Sn Alloys 19
   3.4 Phase Analysis – XRD and EPMA 21
   3.5 Thermal Analysis 21
   3.6 Structure Determination of $Ni_{21}Sn_2P_6$ 23

4. Results in the System Ni-P 31
   4.1 The Ni-rich section between 0 and 35 at.% P 34
   4.2 The central part between 35 and 66.7 at.% P 36
   4.3 The P-rich section with P contents of more than 66.7 at.% 40

5. Results in the System Ni-P-Sn 60
   5.1 Ternary Ni-P-Sn phases 61
   5.2 Phase Equilibria at 850 °C 65
   5.3 Phase Equilibria at 700 and 550 °C 68
   5.4 Thermal Behaviour in the Ni-rich part 78
   5.5 A brief note on the P-Sn system 95
   5.6 Sn-rich phase equilibria 96
   5.7 Conclusion and Lessons Learned from the Ternary Phase Diagram 100

6. The Crystal Structure of C_6Cr_{23}-type $Ni_{21}Sn_2P_6$ (T2) 102
   6.1 An Overview over Selected Ternary Ordered C_6Cr_{23}-type Phases 102
   6.2 Description of the Crystal Structure and Discussion 102
   6.3 Relation to other C_6Cr_{23} compounds 106
   6.4 Comparison of the crystal structures of $Ni_{10}P_3Sn$ and $Ni_{21}Sn_2P_6$ 107

7. Summary 111

   7.1 Summary (English) 111

   7.2 Zusammenfassung (Deutsch) 113

8. References 115
9. Appendices 120

   9.1 List of Figures: 120

   9.2 List of Tables 124

1. Introduction

1.1 Principle of Soldering

Soldering has been known to man for several thousand years. Although the methods have changed significantly and art has matured to technology, the principle of soldering has not: the joining of two higher melting metals (called substrate, contact material or metallization) by means of a third, lower melting one (called solder) [1]. Thus soldering is still the most simple and economical way of joining metal parts.

The process involves three basic steps:
1. melting of the solder and wetting of the substrate
2. diffusion via the interface solder / substrate and chemical reaction between the two
3. solidification of the joint

Step 2 is of course the critical one, because during this step the joint is formed and it is therefore influenced by the process parameters. The occurrence of a chemical reaction is particularly important. The reaction products are intermetallic phases or compounds (abbr. IMC), which are usually brittle and thus determine the quality of the joint, e.g. mechanical properties. Thus cracking often occurs within these IMC layers. Understanding and controlling the growth of these IMCs is therefore a highly important issue in soldering.

In principle, many metals and metallic alloys can be used as contact material or solder [1], but not all element combinations are equally convenient; e.g. Cr in the contact material can lead to dewetting of the surface by the solder [1]. Today's solders are usually Sn-based alloys with additions of various elements. Since 1st July 2006 (RoHS [2] directive of the European Union coming into force) lead-free solder alloys have replaced traditional Sn-Pb containing solders in consumer goods. While the Sn-Pb alloys had excellent properties with respect to process and reliability (e.g. T_m=183 °C, good wetting behaviour, etc.), their successors required – and still require – a considerable amount of research. It is now widely accepted that there is no common drop-in replacement for the conventional Sn-Pb solders. The currently favoured alloys are Sn-Ag-Cu (designated SAC solders, e.g. in Europe Sn 3.8Ag 0.7Cu, amounts in wt.%, in the US Sn 3.9Ag 0.6Cu [3], in Japan Sn 3.0Ag 0.5 Cu [4]) and to a lesser degree Cu-Ni-Sn. Attempts to further optimize the properties of lead-free solders and to lower their melting point have led to the development of Sn-Zn based solders or RE (rare earth) containing solders. These solders, however, suffer from either corrosion proneness (e.g. Sn-Zn), high price or a high number of elements in the alloy (unfavourable for the industry). For high temperature soldering (operational temperature around > 280 °C) so far no replacement has been found.

The most common contact material is Cu, which cannot be used in pure form due to oxidation during storage. In order to protect the Cu-surface from air and oxidation, surface protection is necessary. The various surface finishes include tin-coating, organic surface protection (OSP) and Ni-coating[1] among many others. Ni is advantageous, because it also acts as a diffusion barrier between solder and Cu. The Ni itself is coated with a very thin Au-layer, which gets completely dissolved during soldering[2], whereas various IMCs can be formed by reaction of solder and Ni.

While the formation of the IMC layer(s) is necessary in order to form a good solder joint, it should not grow too thick, as it may then become detrimental to the joint in terms of mechanical stability. It is not surprising that quite an amount of literature deals with this aspect of soldering. Practically, profound knowledge of the process parameters is essential.

1.2 The Influence of Process Parameters with the Focus on Lead-Free Soldering

The soldering process is governed by the interplay of a high number of material properties and parameters. A few of these are shown in Figure 1.1.

Properties and Process Parameters

- melting point / range
- alloy composition
- flux
- atmosphere
- etc.

- process temperature
- reaction products (brittle IMCs)
- wetting
- etc.

Fig. 1.1: Selected properties and parameters influencing the soldering process

Out of the many parameters, the melting point – or rather melting range – is the most obvious criterion. The melting has to occur in a temperature range that is high enough to enable the chemical reaction between solder and contact material, but should be low enough so that components are not destroyed. Flux and atmosphere are equally important, as they influence the wetting behaviour. Fluxes are rosin based, organic substances used to remove oxides from the surface and enhance wetting [1].

Oxides in general are known to have particular devastating effects, among them the so-called "black-pad" phenomenon on PCB boards coated with P-containing Ni layers (see Chapter 1.5).

---

[1] Depending on the plating process this Ni-layer is not necessarily pure, as in many applications a P-containing Ni-layer is used – see Chapter 1.5
[2] Sn-Zn solders have a different behaviour: as Au and Zn form intermetallic compounds, the Au-layer is not dissolved, but takes part in the interfacial reaction resulting in the appearance of Au-Zn intermetallic compounds.

Corrosion of this Ni(P) layer due to poor manufacturing is one of the root causes of this failure, because the corroded layer cannot react with the solder [5]. The particular danger of this phenomenon is that it is not obvious, because the failed joint does not look different from a proper one, and the device thus can easily pass quality control.

Another important parameter is the soldering time, which needs to be set up properly, because growth and thickness of the IMC layer also depend on the soldering (i.e. reaction) time. Complex boards with large components impose their own problems because of unequal heating.

Due to the complex relations among these parameters, change of one parameter will result in the necessity to change others accordingly. While the replacement of Pb by other elements seemed trivial at first sight, this was (and still is) indeed not the case. For SAC alloys the melting range is around 220 °C, which is more than 35 °C higher than the melting temperature of (near) eutectic Sn-Pb alloys (T_m = 183 °C). This higher temperature has direct influence on the use of components, which have to be able to sustain the higher temperature. Furthermore, organic fluxes have to be replaced in order to avoid decomposition before the operating temperature is reached.

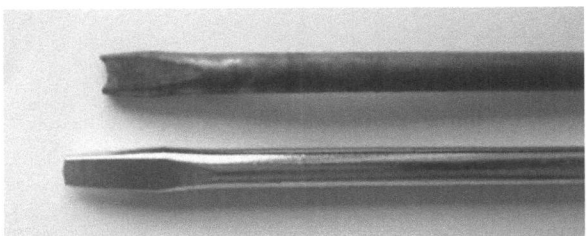

Fig. 1.2: Image of a solder tip of a commercial 20W soldering iron destroyed by use with Sn 3.8Ag 0.7Cu solder. A comparison with the length of a new tip shows the amount of material dissolved during soldering. There was no mechanical influence in the destruction of the tip.

Another detrimental effect of SAC-solders is the dissolution of parts of the equipment that come into direct contact with the solder. Fig. 1.2 shows the tip of a commercial soldering iron destroyed by dissolution of tip-material by the solder, compared to a new tip. This is of course an inconvenience for the user, but is a severe problem for the industry who faces in principle the same effect with their solder pots. In order to avoid a complete change of equipment the use of Cu-Ni-Sn solders has been suggested for wave soldering, because they do not show this effect.

1.3 Contemporary Soldering Techniques

Since its origins in the area of today's country of Afghanistan [1], the soldering process has changed almost beyond recognition. The electronics industry, who faced the need to join many components onto a printed circuit board (PCB), in the first place was the driving force behind the development of various contemporary mass soldering techniques, such as wave soldering or reflow soldering. The current state of the art is *surface mount technology* (SMT), where components are first glued to the surface of the board and then have their leads soldered to the contact material using solder paste in a reflow process. This fabrication allows for high production rates at reasonable cost. However, it is rather demanding in terms of reproducibility and process reliability.

Although through hole technology and hand soldering have long been phased out, they still have their place in the production chain, mainly for large and heavy components that need to be separately soldered onto the board, or for repair work.

Whenever a high lead density is required, a *ball grid array* (BGA) is used. This arrangement does not only provide electrical, but also mechanical connection. For this reason, the joint quality in BGA assemblies is particularly important. The principle of a BGA setup can be seen in Fig. 1.3.

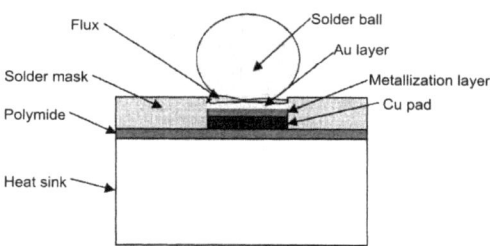

Fig. 1.3: Solder ball attachment on a BGA substrate (from Ref. [6]).

Copyright (2005); reprinted from the cited source with permission from Elsevier

While most process related issues concerning the lead-free transition have been solved in the mean time, the focus has now shifted towards reliability, because the industry lacks knowledge on the behaviour of the solder joints during the lifetime of the device. This behaviour is once again determined by properties of the IMC layer (thickness, composition of the phases, brittleness, holes, etc.) formed within the joint or by production failures (e.g. black-pad phenomenon, Refs. [5] and [7]). The interactions between solder and contact material during soldering therefore become the central issue. Basic knowledge of various properties of an intermetallic alloy system is provided by its phase diagram. Furthermore, the phase diagram is a convenient tool for the interpretation and understanding of interactions between individual phases and the reactive formation of phases.

1.4 Phase Diagrams and (Lead-Free) Soldering

The phase diagram, which is a concise graphical representation of the equilibrium conditions among various phases with respect to composition, temperature and in principle many more parameters, has the same meaning to a researcher as a map has to a traveller. It provides the scientific basis for the development of tailor-made materials and processes ("materials design"). The most prominent example of a technologically important phase diagram is the metastable $Fe - Fe_3C$ phase diagram, which is the foundation of all modern steel technology. Phase diagrams are equally important in the selection process of materials for nuclear fission reactors, where they give critical information whether combinations of materials are stable at the operating temperature or will react to form a lower melting alloy (cause for nuclear meltdown) [8].

In case of soldering, the phases (compounds) in the IMC layer, i.e. the reaction products formed between solder and contact material, are related to the sequence of phases depicted in the phase diagram, because both depend on the same set of thermodynamic parameters. As soldering is a diffusion controlled process, in which two alloys react with each other, many different phases can appear in a solder joint. This is due to the fact that the diffusion path, which indicates the sequence of intermetallic phases formed in a diffusion couple (e.g. a solder joint), has to obey the law of mass conservation. It therefore has to cross the direct connection between the two end-members[3] drawn in a ternary phase diagram (the so called mass balance line).

Of course the solidification behaviour, too, can be understood from the phase diagram. Phase diagrams of systems "solder + contact material" are therefore particularly important, because they can help to understand and interpret phenomena occurring in solder joints. This knowledge forms the basis for the whole design and applicability of the soldering process from early alloy selection to everyday reliability criteria (see schematic illustration in Fig. 1.4).

The investigation of systems "solder + substrate" usually involves the metals Sn and a selection from Cu, Ni, Ag, Au, Pd, etc. For example, recently the quaternary phase diagram Ag-Cu-Ni-Sn and its binary and ternary sub-systems were investigated within an Austrian Research Fund Project (P16495 – N11), because they allow the understanding of interactions between SAC-solders and Ni-substrate.

Generally, the systems "solder + substrate" are characterized by a huge difference in the melting points of the pure metals. This causes a number of experimental problems, e.g. at most temperatures

---

[3] End-members: the two alloys or metals being brought into contact in a diffusion couple; in this case these would be solder and base metal

the system is partially liquid, or at lower temperatures the higher melting regions are experimentally inaccessible due to slow diffusion resulting in non-equilibrium conditions. These difficulties can be overcome by extrapolating the phase equilibria from higher temperatures and / or by modelling the system based on thermodynamic parameters (CALPHAD, CALculation of PHAse Diagrams). In both cases a consistent experimental description of the phase diagram and its thermodynamic properties is required. Thus not only the Sn-rich region of such a system (as it is frequently claimed) needs to be investigated, but in fact the whole system at various temperatures. Furthermore, knowledge of thermodynamic properties allows the verification of the experimental work and its thermodynamic viability by CALPHAD, the modelling of areas inaccessible to experiments and also the prediction of other properties, e.g. surface tension and the wetting behaviour.

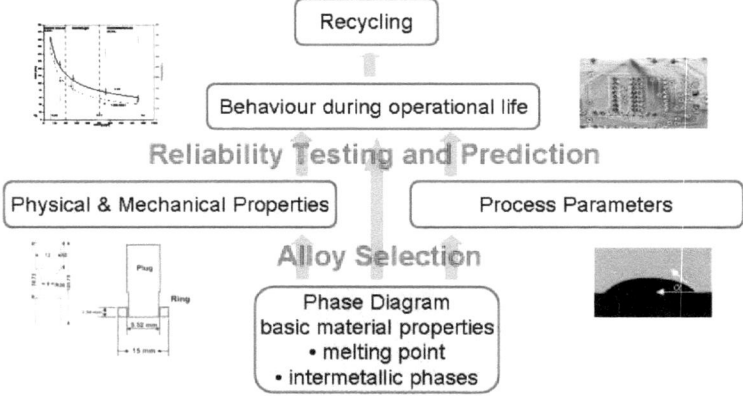

Fig. 1.4: Schematic showing the influence of the phase diagram on the whole soldering process from alloy selection via testing of the method to the end of life treatment.

1.5 The Role of the Ni-P-Sn Phase Diagram

The Ni-coating usually applied on the Cu-surface (see Chapter 1.2) of *under bump metallization* (UBM) in BGA assemblies contains a significant amount of phosphorus when prepared by a process called electroless plating. This method involves the autocatalytic deposition of Ni from a bath containing a solution of $NiSO_4$, NaH_2PO_2 (sodium hypophosphite), and a complexing agent used as a pH-buffer and to prevent precipitates [9]. While this process is simple from a practical point of view, the co-deposition of phosphorus cannot be avoided so that a Ni(P) layer is obtained. The mechanism of this process has not yet been fully clarified.

The deposition rates and P-content depend on the composition of the bath and on the complexing agent: they range from 25.89 µm / hour and 7.39 wt.% P (acetic acid as complexing agent) to 3.97 µm /hour and 11.49 wt.% P (aspartic acid as complexing agent)[4]. A P-content higher than 9 at.% [9], 9.5 at.% [10] or 12.5 at.% [11] will lead to the formation of an amorphous layer.

The UBM is completed by the final Au-layer on top of the Ni(P) layer, and the final assembly is thus called *electroless nickel – gold* (ENIG). As it has been outlined in Chapter 1.2, the Au-layer gets entirely dissolved during soldering and is thus seldom considered in solder related research. It will therefore not be considered in the present work, either.

The wide spread use of Ni(P) coatings is due to the apparent ease of their preparation, the uniform thickness of the layer and good wetting characteristics [12]. However, according to Islam et al. [13] these Ni(P) layers are slightly inferior to electrolytic Ni layers with respect to joint shear strength, especially after long-time reflow.

In terms of soldering the knowledge and control of the interactions between such a Ni(P) layer and Sn-based solders are of particular interest. For the use of high-Sn solders there is in general a high risk of excessive IMC growth, fast consumption of the metallization and spalling of IMCs, so that slowing down reactions between solder and substrate has become one of the prime concerns [14]. Furthermore, the use of phosphorus in solders themselves is not uncommon, too, as its addition enhances the corrosion resistance during the soldering process. A number of P-containing solders have been patented: e.g. Sn-Cu-Ag-Ni-P [15], Sn-Cu-Ni-P [16].

It is therefore not surprising that the reactive phase formation between lead-free solders and Ni(P) substrates has been the subject of a high number of studies[5]. Reaction of the Ni(P) layer and the solder can result in the formation of various IMCs at the interface: Ni_3Sn_4 (e.g. Ref. [11], [14] and [17-19]) if no Cu is involved, $(Cu,Ni)_6Sn_5$ (for $\geq$ 0.7 wt.% Cu in the solder [20],[21]) and $(Cu,Ni)_3Sn_4$ (for 0.5 [20] or 0.2 [21] wt.% Cu in the solder) or both (for 0.5 wt.% Cu [21]). Similar observations were made by Li et al. [22] for the use of Sn-Bi and Sn-Bi-Cu solders with Ni(P) metallization, respectively. In general, the reaction behaviour on Ni(P) is more complex than on pure Cu-surfaces, that act as a Cu source anyway regardless of the solder composition. On Ni(P) surfaces the IMC formation essentially depends on the composition of the solders, i.e. whether the solder contains Cu and can act as a source for Cu. Of course, the formation of these IMCs is

---

[4] values given for a bath containing 0.1 M $NiSO_4 \cdot 6\ H_2O$, 0.25 M $Na_2H_2PO_2 \cdot H_2O$, 0.6 M complexing agent and 0.1 mg / L Pb (stabiliser)
[5] Due to the high amount of available literature only a very brief summary of some selected articles can be given here. Furthermore the studies performed vary in many details making the comparison even more complicated.

governed by completely different reaction and diffusion kinetics. Reaction kinetics between Ni and Sn are usually very slow [7], whereas $(Cu,Ni)_6Sn_5$ can form thick layers.

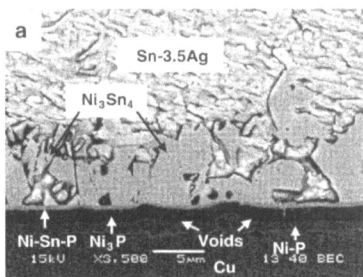

Fig. 1.5: Back scattered SEM image of Sn-3.5Ag/Ni-P/Cu interface directly after reflow soldering at 250 °C for 60s [23]

Copyright (2006); reprinted from the cited source with permission from Elsevier

This interfacial reaction is practically always accompanied by the (partial) transformation of the initial Ni(P) layer into Ni_3P by Ni depletion due to consumption during the reaction via so-called solder reaction-assisted crystallization (e.g. [11, 14, 17-20, 24-26]. This Ni_3P formation can deteriorate the joint quality, because it cracks easily [20], forms channels and thus allows the diffusion of Cu from the underlying board into Ni_3Sn_4 [11] in the interface. On the other hand, Wang and Liu [26] consider the formation of Sn-Cu IMC during the use of Sn-1.0Cu and Sn-3.0Cu solders to be favourable, because they shield the Ni(P) layer from the solder and thus retard the Ni_3P growth. A typical solder joint interface of Sn-3.5Ag/Ni-P(3.9μm)/Cu after reflow soldering at 250 °C for 60 s can be seen in Fig. 1.5 (Ref. [23]).

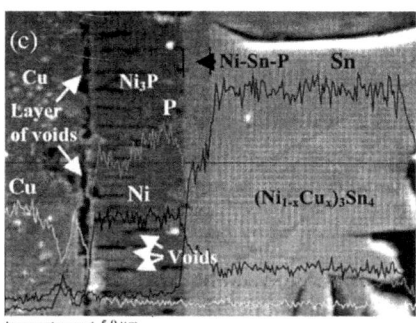

Fig. 1.6: Line scanned SEM image of a Cu/electroless Ni-P/Sn-3.5Ag interface after aging at 200 °C for 48h [11].

Copyright (2006); reprinted from the cited source with permission from Elsevier

Aging of the joint can result in even more complex layers comprising Cu-Ni-Sn (e.g. [24]) and Ni-P-Sn (e.g. Refs. [23, 24, 27]) intermetallics – a typical example is shown in Fig. 1.6 [11]. According to Huang et al. [28] ternary Ni-P-Sn layers were only observed in solder joints of Cu-less solders, which is only partly in agreement with other reports. These reactions together with a change in the volume of the layer also reduce the joint quality [11] because of the build-up of stress in the joint. He et al. [29], too, report a decrease of the strength of Sn-3.5Ag joints due to IMC growth during aging, and the build-up of internal stresses. In case of high P-contents in the original Ni(P) layer, e.g. 13 wt.%, Ni-depletion results in $Ni_{12}P_5$ being formed instead of Ni_3P, followed by the quick formation of Ni_2P and Ni_2PSn [30] after one reflow step. Normally ternary Ni-P-Sn IMC is observed after liquid or solid state aging ([11] and [23]). Spalling of Ni_3Sn_4 from the Ni(P) UBM has been identified as an other detrimental effect, which has been recognized to occur already during the reflow process (Refs. [19, 31, 32]. This effect accelerates the growth of Ni_3P and Ni_3PSn compound layers in the joint, which is a source for brittle fracture. It can result in dewetting of the molten solder, too [31].

Several attempts have been made to decrease the consumption of the Ni(P) layer. Sharif and Chan [33] and Islam et al. [34] found less consumption of Ni(P) by adding In to SAC solder alloys. The addition of Zn, which is said to act as a diffusion block for Ni, to Sn-Ag solders has been reported to reduce the total IMC thickness in the solder joint and to almost entirely suppress the formation of Ni_3PSn [19]. Sharif and Chan [35] even report the total absence of Ni_3P or Ni-P-Sn layers in the joint for the use of Sn-9Zn solders and thus suggest Sn-9Zn / Ni(P) as a good combination.

From the somewhat chaotic picture[6] created by the various literature reports, it can be resumed that the use of Ni(P) coatings is advantageous in many respects, but suffers from two large problems: the spalling of IMC into the solder and brittle fracture. Based on the literature reports given above, the reason for these effects can be directly related to the IMC formation during soldering, i.e. the reaction products of solder and substrate.

However, all these studies lack the knowledge of the phase equilibria in the relevant systems "solder + substrate", e.g. Ni-P-Sn, Cu-Ni-Sn, which is essential for the understanding of IMC formation. This can be seen on an unclear interpretation in the work of Kumar et al. [27], who report the formation of a phase being either similar to a ternary solid solution of P in Ni_3Sn_2 (described by Furuseth and Fjellvåg [36]) or being a mixture of phases on annealing of test joints at 200 °C. As the large solid solution of P in Ni_3Sn_2 exists at 850 °C according to Ref. [36], but

---

[6] Note that many different solders with respect to the element selection and composition were used in the studies. The situation is furthermore complicated by the use of different testing methods which is due to the lack of standard test methods at the time the studies were performed. While common trends can be elucidated from the literature, there is an abundance of variations and differences in the results at detail level.

definitely not below 700 °C, the first interpretation by Kumar et al. has to be wrong and clearly shows the need for a reliable phase diagram description.

The formation of Ni_3PSn in solder joints (which has already been briefly mentioned above) has been reported by many authors in the literature [19, 31, 32, 37]. This phase is claimed to have been described by Furuseth and Fjellvåg in Ref. [36] (see ICSD, Ref. [38]), but is nowhere explicitly mentioned in the original paper (see also Chapter 2). Hwang et al. identified an $InNi_2$-type phase in their solder joints by TEM which they interpreted to be Ni_3PSn. Unfortunately the measured composition of this phase is not given by any author [19, 31, 32, 37] so that this observation remains somewhat inconclusive. However, this fact again highlights the need for a consistent description of the phase equilibria in the ternary system Ni-P-Sn.

For a full understanding of the reaction between SAC solders and Ni(P) substrates the quinary system Ag-Cu-Ni-P-Sn would be needed to be investigated. The representation of such a high order system is not only very complex, but also essentially depends on the quality of the description of its lower order constituents. Indeed, there is a proposal for a group project within COST Action MP0602 on High Temperature Solders dealing with the investigation of this quinary system and its constituents with the focus on the P-containing systems such as Ni-P-Sn, where knowledge is still rather poor.

This ternary system may also gain additional interest with respect to high temperature solders, as the tailored design of solders based on combining proper amounts of Ni-P eutectic and Cu-Sn eutectic pre-alloys has been proposed [39]. Together with knowledge on the ternary Cu-Ni-Sn system [40], which is currently under investigation, too, required information for this materials design process can be provided.

Cu-Ni-P-Sn gets additional importance in flip chip technology, where the diffusion of Cu from the lead of the semiconductor device towards the Ni(P) metallization of the PCB is reported (coupling effect). This can take place within the timescale of the reflow process, and results in the formation of a multi layered IMC interface between solder and metallization [41].

Therefore the aim of the thesis presented here is the investigation of the phase equilibria of the system Ni-P-Sn and its binary constituent systems Ni-P and P-Sn. Knowledge of the phase equilibria of these systems will also be essential for the investigation of higher order systems and for their assessment combining the experimental results and CALPHAD modelling.

This work was funded by the Austrian Research Fund under project No. P18968-N19. It is also meant as a contribution to the European COST Action MP0602 on High Temperature Solders and has already contributed to COST Action 531 (Lead-Free Solders) and ELFNET (European Lead-Free Soldering Network).

2. Literature Review

2.1 The Binary System Ni-Sn

The most recent experimental investigation of this system was done by Schmetterer et al. [42]. Their version of the phase diagram (shown in Fig. 2.1) differs considerably from the last assessment by Nash and Nash [43, 44] and two calculated versions of the phase diagram by Ghosh [45] and Liu et al. [46]. Relevant details of the phases found in the recent study are given in Table 2.1, while the invariant reactions are listed in Table 2.2. A detailed account of all changes in this system can be found in Ref. [42], so that only a short overview is given here.

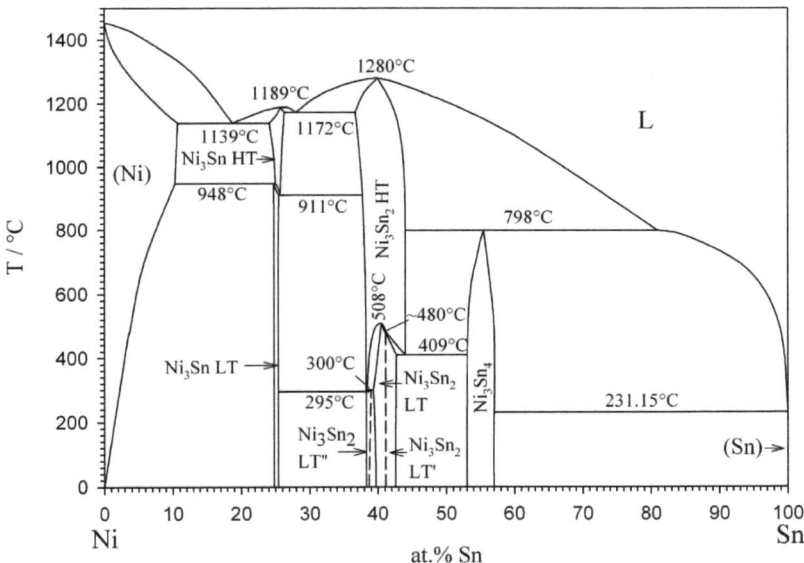

Fig. 2.1: Binary Ni-Sn phase diagram according to Ref. [42].

Copyright (2007); reprinted from the cited source with permission from Elsevier

According to the recent experiments, the transition from the Ni_3Sn low temperature phase (Ni_3Sn LT) to its high temperature modification (Ni_3Sn HT) comprises a eutectoid and a peritectoid reaction in good agreement with earlier results of Havlicek [47]. The determined reaction temperatures of p6 (948 °C) and e7 (911 °C) are in contradiction to most values given in the existing literature, but they are consistent with the evaluations of Mikulas and Thomassen [48] and Havlicek [47]. The Ni_3Sn LT-phase has a narrow homogeneity range of about 0.5 at.%, whereas the homogeneity range of the HT-phase widens at higher temperature. This phase was found to have a cubic BiF$_3$-type structure, as already mentioned earlier by Schubert et al. [49]. An orthorhombic phase, also suggested for the Ni$_3$Sn HT-phase, was confirmed to be metastable, formed by martensitic reaction at high quenching rates.

The region around Ni$_3$Sn$_2$ is far more complicated than previously reported. Besides a NiAs-type Ni$_3$Sn$_2$ HT-phase there are three corresponding orthorhombic LT-phases, two of which have incommensurate structures. Detailed crystal structure analyses in this region were performed by Leineweber et al. [50-53]. The transition between the LT- and HT-phase is of first order, consisting of two eutectoid and two peritectoid reactions.

The homogeneity range of the Ni$_3$Sn$_4$ phase extends from 53 – 57 at.% Sn. In the vicinity of this latter compound several other phases have been reported in literature [47, 54-57], which could not be confirmed in the most recent work [42] and are therefore not included in the phase diagram.

Table 2.1: Solid phases in the binary Ni-Sn system according to Ref. [42]

Phase	Composition Maximum range [at.% Sn]	Pearson symbol	Space group	Strukturbericht designation	Prototype
(Ni)	0 to 10.7	cF4	$Fm\overline{3}m$	A1	Cu
Ni$_3$Sn HT	24.1 to 26.3	cF16	$Fm\overline{3}m$	D0$_3$	BiF$_3$
Ni$_3$Sn LT	24.8 to 25.5	hP8	$P6_3/mmc$	D0$_{19}$	Ni$_3$Sn
Ni$_3$Sn$_2$ HT	36.7 to 44.0	hP6	$P6_3/mmc$		InNi$_2$
Ni$_3$Sn$_2$ LT"	~38.3 to ~39		Cmcm *)		
Ni$_3$Sn$_2$ LT	39.3 to 41.1	oP20	Pnma		Ni$_3$Sn$_2$
Ni$_3$Sn$_2$ LT'	41.25 to 44.0		Cmcm *)		
Ni$_3$Sn$_4$	53.0 to 57.0	mC14	C2/m		Ni$_3$Sn$_4$
(βSn)	~100	tI4	$I4_1/amd$	A5	βSn
(αSn)	~100	cF8	$Fd3m$	A4	αSn
Metastable phases					
Ni$_3$Sn martensite		oP8	Pmmn	D0$_\alpha$	β–Cu$_3$Ti

*) symmetry of average cell with modulation vector α

Table 2.2: Invariant Reactions in the system Ni-Sn according to the literature [42]

Reaction	Designation in this work	Temperature [°C] and Type
L = (Ni) + Ni_3Sn HT	e12	1139, Eutectic
Ni_3Sn HT + (Ni) = Ni_3Sn LT	p6	948, Peritectoid
L = Ni_3Sn HT	melting	1189, Congruent
L = Ni_3Sn HT + Ni_3Sn_2 HT	e13	1172, Eutectic
Ni_3Sn HT = Ni_3Sn LT + Ni_3Sn_2 HT	e7	911, Eutectoid
Ni_3Sn_2 HT = Ni_3Sn_2 LT'' + Ni_3Sn LT	e2	295, Eutectoid
Ni_3Sn_2 HT + Ni_3Sn_2 LT = Ni_3Sn_2 LT''	p1	300, Peritectoid
L = Ni_3Sn_2 HT	melting	1280, Congruent
Ni_3Sn_2 HT = Ni_3Sn_2 LT	congruent transition	508, Congruent
Ni_3Sn_2 HT + Ni_3Sn_2 LT = Ni_3Sn_2 LT'	p2	~ 480, Peritectoid
Ni_3Sn_2 HT = Ni_3Sn_2 LT' + Ni_3Sn_4	e3	409, Eutectoid
Ni_3Sn_2 HT + L = Ni_3Sn_4	p3	798, Peritectic
L = (Sn) + Ni_3Sn_4	e1	231.15*[)], Eutectic

*) value from Refs. [43, 44]

2.2 The Binary System Ni-P

The most recent phase diagram compilation of the Ni-P system was presented by Lee and Nash [44, 58] based on various experimental studies. However, as most of the phase diagram work focused on the Ni-rich part up to about 35 at.% P, this description was far from complete. In the following section the available key information will be summarized.

The early studies by Jolibois [59] and Scholder et al. [60] dealt mainly with preparative aspects of Ni-P alloys; the authors reported the existence of NiP_2 and NiP_3 [59], and Ni_5P_2, Ni_2P and Ni_3P_2[7] [60], respectively. The basic outline of the Ni-rich area was established by Konstantinov [61] based on thermal analysis data obtained from cooling curves and microscopic investigations. From his observations, the author concluded the existence of Ni_3P, Ni_5P_2 and Ni_2P. He also claimed the existence of a low temperature (LT) and a high temperature (HT) modification of Ni_5P_2 based on the observation of the relevant thermal effects. Nowotny and Henglein [62] performed an XRD investigation of this region and produced crystallographic information on unit cell dimensions and crystallographic systems of Ni_3P, Ni_2P, Ni_7P_3[8] and Ni_5P_2. The solubility of P in Ni was established by Koeneman and Metcalfe [63]. A comprehensive study including XRD, DTA and metallography of the Ni-rich part was done by Yupko et al. [64] who confirmed the previous results and introduced

---

[7] Not reported by any other authors, probably equal to Ni_5P_4
[8] Later found to be equal to $Ni_{12}P_5$ [63]

the HT modification of $Ni_{12}P_5$, formed in a peritectic reaction out of Ni_5P_2 HT and the liquid according to their data. The most recent phase diagram study was done by Oryshchyn et al. [65] who investigated several Ni-rich alloys using EPMA and XRD. These authors also determined the crystal structure of the Ni_5P_2 LT phase.

The remaining literature dealing with the Ni-rich region focused on the determination of crystal structures of the various phases: Ni_3P by Aronsson [66] and Rundqvist et al. [67]; Ni_5P_2 LT by Saini et al. [68] (only cell dimensions) and Oryshchyn et al. [65] (full crystal structure); $Ni_{12}P_5$ LT by Rundqvist and Larsson [69]; and Ni_2P by Rundqvist [70]. Therefore the Ni-rich section of the phase diagram can be regarded as reasonably well established although crystal structure information on the HT modifications of Ni_5P_2 and $Ni_{12}P_5$ is still missing.

The central part of the phase diagram between 35 and 66.7 at.% P is based on work by Larsson [71] and has still quite tentative character in the compilation of Lee and Nash [58]. Due to the high vapour pressure of P in this area, the interpretation of experimental results becomes rather difficult. Based on his extensive investigation, Larsson [71] reported the existence of the phases NiP (including crystal structure determination) and $Ni_{1.22}P$, and presented a possible phase diagram for this region. In addition, he mentioned the existence of a metastable eutectic $L = Ni_2P + NiP$. However, since no thermal analyses have been performed in this region, the reported invariant reactions still have rather tentative character. Further work in this central section was done by Elfström who determined the crystal structure of Ni_5P_4 [72].

Information on the P-rich section (with P contents beyond 66.7 at.%) was given by Jolibois [59] and Biltz and Heimbrecht [73]. The latter authors investigated the degradation of P-rich samples due to evaporation of P on heating. They derived the existence and composition of NiP_3, NiP_2 and Ni_2P from the development of the vapor pressure and an analysis of the remaining solid material. For concentrations of 44 to 47 at.% P they reported melting of the alloy at approximately 900 °C. The crystal structures of NiP_2 and NiP_3 were determined by Larsson [71] and Rundqvist and Ersson [74], respectively.

Recently Shim et al. [75] attempted a thermodynamic modeling of the binary Ni-P system. However, their version suffers from an incomplete description of the thermodynamic properties, especially at higher P concentrations. This leads to severe problems with the liquidus line which drops down to 0 K at approximately 45 at.% P in their diagram.

Relevant literature data on the various phases are summarized in Table 4.3, while information on invariant reactions is given in Table 4.4 (in both cases together with results from the present work).

2.3 The Binary System P-Sn

The current version of the P-Sn phase diagram reprinted in Massalski et al. [44] is almost entirely based on the work of Vivian [76]. A further study based on powder XRD is available from Olofsson [77]. However, no phase diagram version was established by this author. Besides the pure elements three binary compounds have been included in the phase diagram: P_3Sn_4 (crystal structure from Ref. [78]), P_4Sn_3 (crystal structure from Ref. [79]) and P_3Sn [80]. Another compound, PSn, has been reported by Katz et al. [81] having a hexagonal unit cell. According to Zaikina et al. [79] P_4Sn_3 does not have a significant homogeneity range, which contradicts the literature phase diagram. Furthermore they report that a certain pressure is required for the formation of PSn.

The phase diagram version established by Vivian [76] is based on the chemical and micrographical analysis of samples prepared in pressure tubes. From the appearance of the samples the author concluded the existence of two liquid miscibility gaps. Characteristic temperatures were obtained from thermal analysis using cooling curves only. The formation of the P_3Sn_4 phase is reported to work according to a syntectic reaction L1 + L2 = P_3Sn_4 at approx. 550 °C and is thus related to the Sn-rich liquid miscibility gap.

An important feature of this system is the high vapour pressure of the samples. According to the observations of Vivian [76] samples having a higher P-content than 8.5 wt.% P cannot be prepared at normal pressure and so-called pressure tubes have to be used instead. These samples are reported to melt under the considerable evaporation of P which contributes to the build up of pressure in the tube. Therefore, the P-Sn phase diagram is definitely not isobaric. Indeed a certain P-vapour pressure appears to be necessary to allow the formation of the reported phases.

2.4. The Ternary System Ni-P-Sn

Phase diagram information in the system Ni-P-Sn is scarce. Most available literature deals with the determination of crystal structures of ternary Ni-P-Sn compounds. A total of four ternary compounds have so far been described in the literature: $Ni_{10}P_3Sn$ (T1) [82], $Ni_{10}P_3Sn_5$ (T3) [83], $Ni_{13}P_3Sn_8$ (T4) [84] and Ni_2PSn (T5) [85]. $Ni_{13}P_3Sn_8$ (T4), $Ni_{10}P_3Sn_5$ (T3) and Ni_2PSn (T5) are related to the NiAs type structure, and indeed the first two are reported to form out of a large ternary solid solution of the Ni_3Sn_2 high temperature (HT) phase on cooling [36]. This ternary solid solution has the general formula $Ni_{1+m}P_xSn_{1-x}$ ($0.00 \leq m \leq 0.65$, $0.00 \leq x \leq 0.32$) according to Ref. [36]. The homogeneity range was determined by the disappearing phase principle and from the unit cell dimensions, whereas no EPMA / EDX measurements were done. Later Furuseth et al. [86] described ordering phenomena of P and Sn within this large ternary solid solution of Ni_3Sn_2 HT based on electron diffraction measurements and proposed a structural model for the composition $Ni_{52.4}P_{14.3}Sn_{33.3}$ based on electron diffraction. The crystal structures of $Ni_{13}P_3Sn_8$ (T4) and $Ni_{10}P_3Sn_5$ (T3) were elucidated using electron diffraction and single crystal XRD, but their formation mechanisms out of Ni_3Sn_2 HT have not been investigated [83, 84].

Ni_2PSn (T5) is reported to decompose into Sn (sic!) and Ni_2P at 732 °C [85]. Its crystal structure shows some similarities to the MnP type, but in the ternary compound there is a complete ordering of P and Sn. $Ni_{10}P_3Sn$ (T1) has its own crystal structure type, and it was claimed to have a melting point at approx. 850 °C [82].

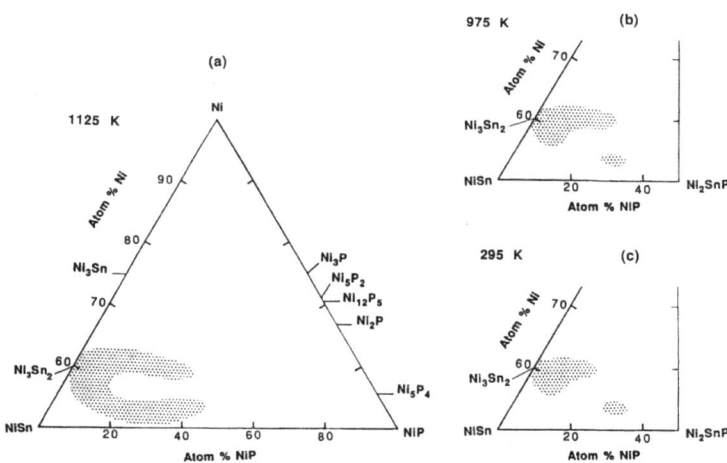

Fig. 2.2: Section of the Ni-P-Sn phase diagram at (a) 1125 K, (b) 975 K and (c) 295 K (judged from slowly cooled samples). [36]

Copyright (1994); reprinted from the cited source with permission from the Danish Chemical Society

A further compound Ni_3PSn is frequently cited in literature related to the IMC evolution in lead-free solder joints [19, 31, 32, 37]. Information on this phase, however, remains questionable. According to the reference given in the Inorganic Chrystal Structure Database (ICSD) [38] Ni_3PSn has been described in the work of Furuseth and Fjellvåg, Ref. [36], but it is never explicitly mentioned in the original publication. Furthermore, the phase composition is given as $Ni_{1.05}P_{0.3}Sn_{0.7}$

($Ni_{51.2}P_{14.6}Sn_{34.1}$), which corresponds more to the stoichiometry Ni_3PSn_2. Although the composition of this phase is said to have been determined (e.g. using EDX) in the solder related literature, the actual composition of this phase in the solder joint is never explicitly mentioned.

No phase diagram information is available except for a sketch of the solid solution of P in Ni_3Sn_2 HT in Ref. [36] – see Fig. 2.2.

3. Experimental Section

3.1 Preparation of Binary Ni-P Alloys

Ni-P alloys were prepared by mixing proper amounts of Ni powders (99.995%, Koch Light Laboratories, UK) and P powders (red P lump, 99.999+%, Alfa Aesar, Germany) or Ni powders and Ni_2P powders (99.5%, Alfa Aesar, Germany). The P powder was prepared from P lump and all elements were stored either in a glove box or in the desiccator, while weighing of the components could be done on air. The powders were thoroughly mixed and subsequently pressed into pellets. Alloying was directly done in evacuated and sealed quartz glass tubes, and no attack of the samples on the quartz glass was observed. Only samples with a P content higher than 65 at.% were placed in an alumina crucible before sealing in quartz glass.

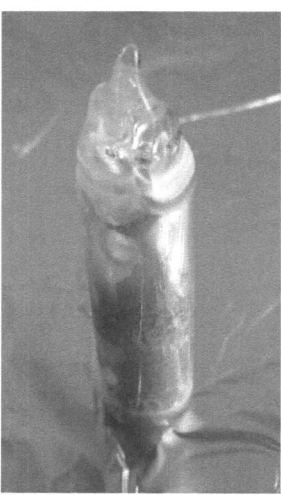

Fig. 3.1: Effect of P-evaporation during annealing: quenching resulted in the condensation of a huge amount of P on the inner quartz glass wall (Ni-P alloy containing 70 at.% P).

According to the literature, the Ni-P system is not really isobaric at P concentrations higher than 40 at.%, i.e. a significant P vapor pressure develops in this region at high temperatures. Furthermore, the reaction between Ni and P is quite vigorous and strongly exothermic: it was found to start rather suddenly at approximately 550 °C resulting in a sudden increase of temperature and pressure in the quartz glass tube which, at the beginning, frequently lead to explosions. These facts severely influenced the heating programmes used during alloying. While in the Ni-rich part the pellets could be heated to 1100 °C within a week without any significant problems, this procedure could not be used for P-rich alloys. After numerous explosions of quartz capsules, the heating program finally chosen required approximately two weeks because a heating rate of 1 K/h had to be used with isothermal segments at 350 °C (12 hours), 450 °C (48 hours; just above the sublimation point of red P) and 550 °C (48 hours). Furthermore, samples in this critical composition range were not heated higher than 700 °C during alloying, because at higher temperatures P evaporation became significant (see Fig. 3.1). For example, samples annealed at 880 °C were found to be hollow – possibly due to the formation of a P bubble within the sample – and a large amount of P condensed on the inner quartz glass surface during quenching (shown in Fig. 3.2). However, the samples annealed at 700 °C for two to three weeks were found to be homogeneous and in equilibrium.

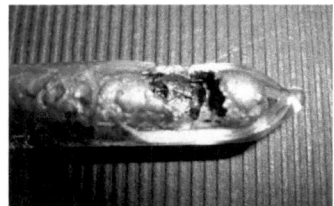

Fig. 3.2: Ni-P sample annealed at 900 °C. The sample is hollow due to evaporation of P.

After annealing the samples were quenched by immersing the quartz tubes into cold water; in some cases the capsules were broken on contact with the water in order to allow higher quenching rates by direct contact of water and sample. The conditions of thermal treatment are given in Table 4.1.

3.2 Preparation of Binary P-Sn Alloys

Binary P-Sn alloys were prepared from P powder made from P-lump (99.999+%, Alfa Aesar, Germany) and Sn-powder (99.999%, Alfa Aesar, Germany). Proper amounts were mixed, pressed into pellets and alloyed in alumina containers that were sealed in evacuated quartz crucibles. For the P-Sn alloys a simpler temperature program compared to the production of Ni-P alloys could be used: the powder mixtures were heated to 350 °C within 12 hours, where they were kept for 48 hours, and were then heated to 500 °C. This temperature was not exceeded in order to avoid excessive evaporation of phosphorus. Equilibrium annealing was carried out at 500 °C. The samples were then quenched in cold water.

3.3 Preparation of Ternary Ni-P-Sn Alloys

The preparation of the ternary alloys in principle followed the method of the binary samples described in the previous Chapters.

Ternary Ni-P-Sn alloys were prepared from powders of Ni (99.995%, Koch Light Laboratories, UK), P (P-lump, 99.999+%, Alfa Aesar, Germany), Ni$_2$P (99.5%, Alfa Aesar, Germany), Ni$_3$Sn and Ni$_3$Sn$_2$ as well as Sn pieces (99.999% Metal Basis, Ventron Alfa Products, USA) or Sn-powder (99.999%, Alfa Aesar, Germany). The powders of Ni$_3$Sn and Ni$_3$Sn$_2$ were made by arc melting, annealing and grinding of the material, followed by a quality check using X-ray diffraction. Again all starting materials were stored either in a glove box or in the desiccator, while weighing of the components could be done on air.

Sample preparation involved mixing and pressing of the powders into pellets and the addition of appropriate amounts of Sn-pieces (if required). All components were filled into an alumina crucible and sealed in evacuated quartz glass tubes. The use of alumina crucibles was found to be necessary, after embrittlement of the quartz glass tube on direct contact with the (liquid) ternary alloy was observed in the first batch of samples. As no Si was found in these samples during EPMA and EDX measurements and the affected samples could easily be removed from the quartz tube, reaction between sample and quartz could be ruled out. The observed embrittlement is assumed to be caused by strong wetting forces between alloy and glass.

After sealing samples with less than 50 at.% P were slowly heated to 1050 °C within one week, where they were kept for six hours followed by equilibrium annealing at 550, 700 or 850 °C. The annealing time depended on the annealing temperature and liquidus temperature estimated from the binary systems and varied between three days and several months (see also Table 5.1).

Similar to the binary Ni-P alloys, samples with more than 50 at.% P had to be heated slowly according to the heating programme described in Chapter 3.1 for the Ni-P alloys.

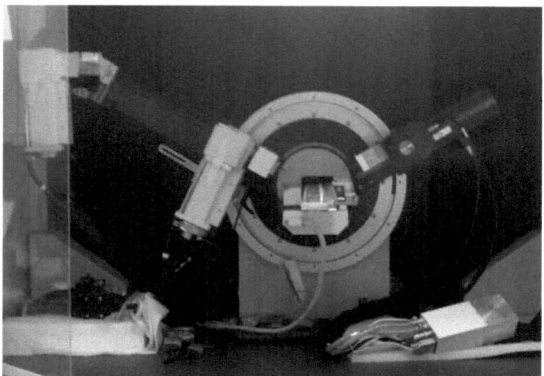

Fig. 3.3: Bruker D8 powder diffractometer used for phase analysis. The X-ray tube is on the left. The autosampler is in the center and the detector on the right. During the measurement the autosampler rotates by the angle θ, while the detector simultaneously rotates by 2θ (Bragg-Brentano geometry).

3.4 Phase Analysis – XRD and EPMA

Powder X-ray diffraction (XRD) data at room temperature were recorded in a Guinier-Huber Image Plate camera (Huber GmbH, Rimsting, Germany) on a Siemens Kristalloflex ERL 1000 generator (Siemens AG, Berlin, Germany) with Cu $K_{\alpha 1}$ radiation. High purity Si powder mixed with the samples was used as an internal standard; exposure times varied between 1 and 3 hours. Alternatively, a number of samples were investigated on a Bruker D8 powder diffractometer equipped with a high speed LynxEye one dimensional silicon strip detector and a Cu X-ray tube. In both cases the patterns were analyzed using the 'Topas 3' software (Bruker AXS, Karlsruhe, Germany). Crystal structure data for phase identification were taken from Pearson's Handbook of Intermetallic Phases [87] or from the original literature as given in Chapter 2 (e.g. Refs. [51, 53, 65]).

Samples to be examined by optical light microscopy and scanning electron microscopy (SEM) were embedded in a phenolic resin mixed with carbon (Struers, Denmark) or a mixture of Resinar F and Cu-powder (volume ratio of 2:1). After embedding, the samples were ground with SiC discs with 120, 240, 400, 800 and 1200 mesh and polished with Al_2O_3 powder (1µm) to obtain a smooth surface.

Metallographic investigations were performed using a light microscope (Zeiss Axiotech, Jena, Germany). A DSC-S75 digital still camera (Sony, Tokyo, Japan) switched to full zoom was employed to take pictures. SEM and electron probe micro analysis (EPMA) measurements were carried out on a Cameca SX 100 instrument (Cameca, Paris, France) using wavelength dispersive spectroscopy (WDS) for quantitative analyses and employing pure Ni and Sn as well as Apatite as standard materials. The measurements were carried out at 20 kV with a beam current of 20 nA. Ni K_α, P K_α and Sn L_α lines were used for quantitative analyses. Conventional ZAF matrix correction was used to calculate the compositions from the measured X-ray intensities.

A number of samples were investigated in cooperation with the Technical University Eindhoven (JEOL JSM-840A, Energy Dispersive X-ray Spectroscopy (EDX); 20 kV / 1-1.5 nA beam current; ZAF correction; calibration via pure Ni, Sn and GaP) and the Institute of Physics of the Czech Academy of Sciences (JEOL JSM-6460, EDX; 15 kV; calibration via Ni, Sn and InP).

3.5 Thermal Analysis

Pieces of the annealed samples weighing 200 to 300 mg were used for differential thermal analysis (DTA) measurements in evacuated and sealed quartz glass crucibles in a Netzsch DTA furnace (Netzsch, Selb, Germany) with a home made controller system using Eurotherm components (Controller No. 3508, Eurotherm, Vienna, Austria). A sketch of the thermocouple and an image can be seen in Fig. 3.4. The use of closed quartz crucibles was necessary in order to avoid the excessive evaporation of P during the measurement, but at the same time limited the highest accessible temperature to 1200 °C.

The measurements were performed at various scanning rates from 0.1 to 5 K/min. Generally two heating and two cooling curves were recorded for each sample. The DTA instrument was calibrated using the high purity metals Sn, Sb and Au as standards to establish an internal calibration file. The reproducibility of the DTA analyses was determined from the melting temperatures of pure metals observed during the three[9] heating / cooling cycles, and it was found to be within 1 °C. The total experimental error, however, which was derived from the scatter of several sample measurements for an invariant effect, was estimated to be less than ±2 °C.

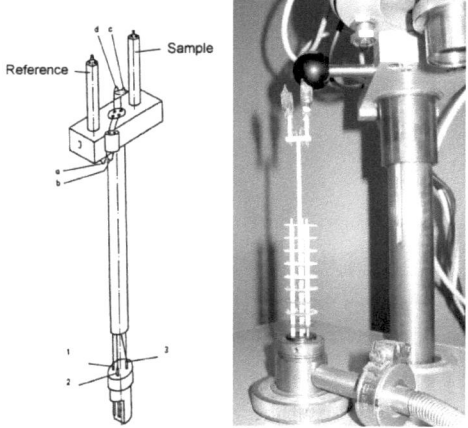

Fig. 3.4: Drawing and image of the experimental DTA setup used in the present study.

Due to the vapour pressure of P the DTA measurements of samples with a high P-content were prone to a number of difficulties: P-loss caused by evaporation changed the sample composition and resulted in the development of a considerable vapour pressure. This influenced the phase equilibria and thus the observed thermal effects. Furthermore, this vapour pressure could lead to the explosion of the quartz crucible when too high, accompanied by the consumption of the Pt-wires of the thermocouples due to chemical reaction with the gas phase (see Fig. 3.5). This, of course, put a limit to the experimental work with high P-containing samples.

---

[9] In case of metal standards three heating / cooling cycles were recorded.

Experimental Section - 23 -

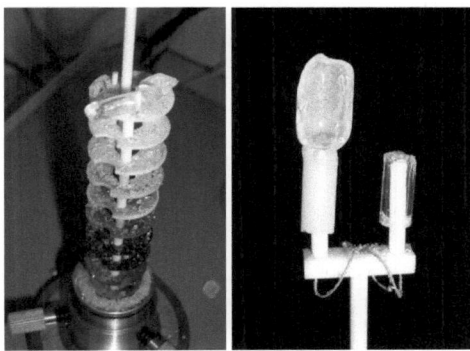

Fig. 3.5: Destroyed transducer unit of the DTA instrument after explosion of a Ni-P sample.

On the other hand, embrittlement of the quartz tubes as observed during initial alloying of the samples (cf. Chapter 3.3) was not (or at least not as strongly) encountered during the DTA measurements. In all cases the crucibles kept tight and did not fall apart during removal from the DTA furnace.

3.6 Structure Determination of $Ni_{21}Sn_2P_6$

Initial sample characterization was carried out at room temperature on the Guinier type system or on the Diffractometer described in Chapter 3.3. For single crystal structure determination of $Ni_{21}Sn_2P_6$ a crystal was isolated from a sample containing the three phases Ni_3Sn_2 high temperature (HT), Ni_3Sn low temperature (LT) and $Ni_{21}Sn_2P_6$ (major phase). The X-ray measurement was performed at room temperature on a Nonius KappaCCD diffractometer equipped with a monocapillary optics collimator (graphite monochromatized MoK$_\alpha$ radiation). Crystal data as well as experimental details are compiled in Table 6.1 (page 103). The measured intensities were corrected for Lorentz, background and polarisation effects as well as for absorption by evaluation of multi-scans. The systematically absent reflections suggested the space groups $Fm\overline{3}m$ and $F\overline{3}m$, of which the first was chosen and confirmed during the refinement. The structure itself was solved using an automatic Patterson method (SHELXS-97 [88]), as the employment of direct methods did not yield any useful results. The structure refinement by full-matrix least-squares techniques on F^2 was done with SHELXL-97 [88]. A single phase sample (No. T2) at the nominal composition of the new phase was finally used for characterization by powder XRD following the method described in Chapter 3.3.

Table 4.1: Experimental results of the phase analysis in the system Ni-P

No.	Nominal Composition [at.%]	Heat Treatment [°C]	Phase	Structure Type	Lattice Param. [pm]	WDS Ni[at.%]	WDS P[at.%]	Σ mass%
NP 2	$Ni_{83}P_{17}$	700, 36 d	(Ni)	Cu	a=352.433(5)	99.3	0.7	98.7
			Ni_3P	Ni_3P	a=895.42(2) c=438.68(1)	75.0	25.0	100.1
NP 7	$Ni_{82}P_{18}$	700, 83 d	(Ni)	Cu	a=352.490(5)	Not determined		
			Ni_3P	Ni_3P	a=895.52(1) c=438.75(1)			
NP 8	$Ni_{81}P_{19}$	700, 36 d	(Ni)	Cu	a=352.346(6)	Not determined		
			Ni_3P	Ni_3P	a=895.56(2) c=438.82(1)			
NP 9	$Ni_{80}P_{20}$	700, 36 d	(Ni)	Cu	a=352.399(6)	Not determined		
			Ni_3P	Ni_3P	a=895.47(2) c=438.77(1)			
NP 3	$Ni_{79}P_{21}$	700, 36 d	(Ni)	Cu	a=352.403(7)	98.9	1.1	99.9
			Ni_3P	Ni_3P	a=895.51(2) c=438.82(1)	74.9	25.1	100.1
NP 10	$Ni_{76}P_{24}$	700, 20 d	(Ni)	Cu	a=352.64(3)			
			Ni_3P	Ni_3P	a=895.686(8) c=438.876(4)			
NP 11	$Ni_{74}P_{26}$	700, 36 d	Ni_3P	Ni_3P	a=895.54(2) c=438.79(8)	75.0	25.0	100.4
			Ni_5P_2 LT	Ni_5P_2	a=661.15(4) c=1231.11(1)	71.8	28.2	100.2
NP 1	$Ni_{73}P_{27}$	700, 36 d	Ni_3P	Ni_3P	a=895.56(2) c=438.75(1)			100.3
			Ni_5P_2 LT	Ni_5P_2	a=661.24(2) c=1232.53(6)	71.8	28.2	
NP 1	$Ni_{73}P_{27}$	1050, 14 d	Ni_3P	Ni_3P	a=895.3(1) 438.80(7)	75.0	25.0	100.0
			Ni_5P_2 LT	Ni_5P_2	a=660.70(8) c=1232.9(2)	71.9	28.1	100.2
NP 12	$Ni_{72}P_{28}$	700, 36 d	Ni_3P	Ni_3P	a=880.88(9) c=450.05(7)	not found in EPMA		
			Ni_5P_2 LT	Ni_5P_2	a=661.14(3) c=1232.16(5)	71.8	28.2	100.5
NP 12	$Ni_{72}P_{28}$	1050, 7 d	Ni_3P	Ni_3P	a=885.0(3) c=445.42(2)	not found in EPMA		
			Ni_5P_2 LT	Ni_5P_2	a=660.66(6) c=1231.8(1)	71.8	28.2	100.2
NP 4	$Ni_{71}P_{29}$	700, 36 d	Ni_5P_2 LT	Ni_5P_2	a=660.83(1) c=1231.86(4)	71.6	28.4	100.4
			$Ni_{12}P_5$ LT	$Ni_{12}P_5$	a=864.67(1) c=507.132(9)	70.7	29.3	100.6
NP 4	$Ni_{71}P_{29}$	1050, 14 d	Ni_5P_2 LT	Ni_5P_2	a=660.56(2) c=1230.84(4)	71.8	28.2	101.0
			$Ni_{12}P_5$ LT	$Ni_{12}P_5$	a=864.50(1) c=507.00(1)	70.8	29.2	100.2
NP 37	$Ni_{71}P_{29}$	1100, 27 d	$Ni_{12}P_5$ LT	$Ni_{12}P_5$	a=864.71(2) c=507.13(3) + strong additional lines	Not determined		
NP 29	$Ni_{70.5}P_{29.5}$	700, 30 d	$Ni_{12}P_5$ LT	$Ni_{12}P_5$	a=864.730(4) c=507.127(3)	Not determined		
NP 29	$Ni_{70.5}P_{29.5}$	DTA cooled	$Ni_{12}P_5$ LT	$Ni_{12}P_5$	a=864.640(8) c=507.062(5)	Not determined		
NP 39	$Ni_{69.5}P_{30.5}$	700, 15 d	$Ni_{12}P_5$ LT	$Ni_{12}P_5$	a=864.667(7) c=507.117(5)	Not determined		
			Ni_2P	Ni_2P	a=586.44(2) c=393.11(2)			
NP 5	$Ni_{69}P_{31}$	700, 36 d	$Ni_{12}P_5$ LT	$Ni_{12}P_5$	a=864.574(4) c=507.13(3)	70.8	29.2	100.5
			Ni_2P	Ni_2P	a=586.08(2) c=339.82(1)	66.9	33.1	100.3

Sample	Composition	Conditions	Phases	Lattice parameters			
NP 5	Ni$_{69}$P$_{31}$	1050, 14 d	Ni$_{12}$P$_5$ LT Ni$_2$P	a=864.619(8) c=507.138(5) a=586.51(1) c=339.15(1)	70.7 66.8	29.3 33.2	100.8 100.6
NP 5	Ni$_{69}$P$_{31}$	DTA cooled	Ni$_{12}$P$_5$ LT Ni$_2$P	a=864.81(1) c=507.27(1) a=586.64(3) c=339.16(3)		Not determined	
NP 13	Ni$_{68}$P$_{32}$	700, 36 d	Ni$_{12}$P$_5$ LT Ni$_2$P	a=864.46(15) c=506.76(10) a=586.32(1) c=338.86(1)	67.5	32.5	100.1
NP 13	Ni$_{68}$P$_{32}$	850, 14 d	Ni$_{12}$P$_5$ LT Ni$_2$P	a=864.51(1) c=506.77(1) a=586.179(9) c=338.849(7)		Not determined	
NP 13	Ni$_{68}$P$_{32}$	1050, 7 d	Ni$_{12}$P$_5$ LT Ni$_2$P	a=864.69(1) c=507.120(8) a=586.39(1) c=338.90(1)	71.0 66.8	29.0 33.2	99.6 100.4
NP 13	Ni$_{68}$P$_{32}$	DTA cooled	Ni$_{12}$P$_5$ LT Ni$_2$P	a=585.92(3) c=339.61(2) a=864.83(1) c=507.20(1)		Not determined	
NP 30	Ni$_{67.2}$P$_{32.8}$	700, 30d	Ni$_{12}$P$_5$ LT Ni$_2$P	a=864.72(4) c=507.22(3) a=586.506(7) c=338.952(5)		Not determined	
NP 30	Ni$_{67.2}$P$_{32.8}$	DTA cooled	Ni$_{12}$P$_5$ LT Ni$_2$P	a=865.0(1) c=506.86(9) a=586.50(2) c=338.28(1)		Not determined	
NP 31	Ni$_{65.5}$P$_{34.5}$	700, 30 d	Ni$_2$P Ni$_5$P$_4$	a=586.133(7) c=338.510(6) a=679.22(3) c=1098.45(8)		Not determined	
NP 14	Ni$_{63}$P$_{37}$	700, 36 d	Ni$_2$P Ni$_5$P$_4$	a=585.99(1) c=338.45(1) a=678.99(1) c=1098.58(2)	66.4 55.9	33.6 44.1	100.4 100.2
NP 17	Ni$_{60}$P$_{40}$	550, 59 d	Ni$_2$P Ni$_5$P$_4$	a=568.65(2) c=338.47(2) a=679.194(5) c=1098.551(9)	66.5 55.8	33.5 44.2	100.5 100.5
NP 17	Ni$_{60}$P$_{40}$	600, 75 d	Ni$_2$P Ni$_5$P$_4$	a=586.68(3) c=338.36(3) a=679.103(4) c=1098.654(8)	66.4 55.9	33.6 44.1	100.4 100.6
NP 6	Ni$_{60}$P$_{40}$	700, 36 d	Ni$_2$P Ni$_5$P$_4$	a=585.77(3) c=339.05(3) a=678.95(5) c=1098.43(10)	66.2 55.9	33.8 44.1	100.0 99.8
NP 17	Ni$_{60}$P$_{40}$	810, 10 d	Ni$_2$P Ni$_5$P$_4$	a=586.28(2) c=338.15(2) a=679.046(6) c=1098.816(9)	66.1 55.7	33.9 44.3	100.4 99.8
NP 21	Ni$_{58}$P$_{42}$	700, 20 d	Ni$_2$P Ni$_5$P$_4$	a=586.72(5) c=337.91(5) a=679.274(6) c=1099.04(1)	66.1 55.9	33.9 44.1	99.8 100.1
NP 21	Ni$_{58}$P$_{42}$	900, 13 d	Ni$_2$P Ni$_5$P$_4$ NiP (in fine eutectic)	a=586.28(3) c=338.46(2) a=678.96(2) c=1098.52(4) a=610.34(4) b=508.35(5) c=666.38(6)		Not determined	

No.	Nominal Composition [at %]	Heat Treatment [°C]	Phase	Structure Type	Lattice Param. [pm]	WDS Ni[at.%]	WDS P[at.%]	Σ mass%
NP 22	Ni$_{57}$P$_{43}$	700, 20 d	Ni$_2$P Ni$_5$P$_4$	Ni$_2$P Ni$_5$P$_4$	a=587.06(3) c=338.08(3) a=679.190(4) c=1098.808(8)	Not determined		
NP 22	Ni$_{57}$P$_{43}$	900, 13 d	Ni$_2$P Ni$_5$P$_4$ NiP (in fine eutectic)	Ni$_2$P Ni$_5$P$_4$ NiP	a=586.14(3) c=338.40(2) a=679.00(2) c=1098.52(3) a=610.12(8) b=506.09(8) c=671.03(9)	Not determined		
NP 16	Ni$_{56}$P$_{44}$	700, 20 d	Ni$_2$P Ni$_5$P$_4$	Ni$_2$P Ni$_5$P$_4$	a=588.1(1) c=337.4(1) a=679.138(4) c=1098.686(8)	Not determined		
NP 16	Ni$_{56}$P$_{44}$	900, 13 d	Ni$_2$P Ni$_5$P$_4$ NiP (in fine eutectic)	Ni$_2$P Ni$_5$P$_4$ NiP	a=586.28(4) c=338.58(3) a=679.06(1) c=1098.52(3) a=610.15(8) b=505.95(7) c=671.58(9)	66.5 56.3 50.3	33.5 43.7 49.7	99.9 100.2 100.0
NP 23	Ni$_{53.5}$P$_{46.5}$	700, 34 d	Ni$_5$P$_4$ NiP$_2$	Ni$_5$P$_4$ NiP$_2$	a=679.330(5) c=1098.661(9) a=635.41(9) b=561.81(7) c=606.93(8) β=126.147(8)°	Not determined		
NP 24	Ni$_{52.5}$P$_{47.5}$	700, 34 d	Ni$_5$P$_4$ NiP$_2$	Ni$_5$P$_4$ NiP$_2$	a=679.403(5) c=1098.71(1) a=635.54(3) b=561.71(2) c=606.99(3) β=126.158(3)°	Not determined		
NP 25	Ni$_{51.5}$P$_{48.5}$	700, 34 d	Ni$_5$P$_4$ NiP$_2$	Ni$_5$P$_4$ NiP$_2$	a=679.383(1) c=1098.526(3) a=635.09(5) b=561.63(4) c=606.96(5) β=126.121(5)°	Not determined		
NP 15	Ni$_{55}$P$_{45}$	700, 36 d	Ni$_5$P$_4$ NiP$_2$	Ni$_5$P$_4$ NiP$_2$	a=679.09(1) c=1098.18(1) a=635.73(14) b=561.25(12) c=606.2(1) β=126.12(1)°	55.8 34.7	44.2 65.3	99.9 99.4
NP 35	Ni$_{51}$P$_{49}$	700, 22 d	Ni$_5$P$_4$ NiP$_2$	Ni$_5$P$_4$ NiP$_2$	a=679.248(5) c=1098.48(1) a=635.33(2) b=561.61(2) c=606.72(2) β=126.135(2) °	Not determined		
NP 18	Ni$_{50}$P$_{50}$	700, 17 d	Ni$_5$P$_4$ NiP$_2$	Ni$_5$P$_4$ NiP$_2$	a=679.695(6) c=1098.18(1) a=635.86(1) b=561.47(1) c=607.00(1) β=126.188(1)°	56.0 34.4	44.0 65.6	100.5 99.7
NP 18	Ni$_{50}$P$_{50}$	810, 10 d	Ni$_5$P$_4$ NiP$_2$	Ni$_5$P$_4$ NiP$_2$	a=679.260(3) c=1098.617(6) not found in XRD	Not determined		
NP 36	Ni$_{49}$P$_{51}$	700, 22 d	Ni$_5$P$_4$ NiP$_2$	Ni$_5$P$_4$ NiP$_2$	a=679.223(5) c=1098.43(1) a=635.65(1) b=561.56(1) c=607.05(1) β=126.174(1) °	Not determined		
NP 26	Ni$_{47.5}$P$_{52.5}$	700, 30 d	Ni$_5$P$_4$ NiP$_2$	Ni$_5$P$_4$ NiP$_2$	a=679.383(1) c=1098.526(3) a=635.09(5) b=561.63(4) c=606.96(5) β=126.121(5)°	Not determined		

NP	Comp.	Cond.	Phases	Lattice parameters	
NP 26	$Ni_{47.5}P_{52.5}$	880, 3 d	Ni_2P Ni_5P_4 NiP NiP_2	$a=586.32(5)$ $c=339.34(4)$ $a=678.93(7)$ $c=1099.5(2)$ $a=605.17(1)$ $b=488.45(1)$ $c=689.64(2)$ $a=641.04(6)$ $b=560.38(6)$ $c=607.62(5)$ $\beta=125.632(6)°$	Not determined
NP 19	$Ni_{45}P_{55}$	700, 14 d	Ni_5P_4 NiP_2	$a=679.304(6)$ $c=1098.55(1)$ $a=636.524(7)$ $b=561.686(6)$ $c=607.270(7)$ $\beta=126.2171(6)°$	Not determined
NP 19	$Ni_{45}P_{55}$	810, 34 d	Ni_5P_4 NiP_2	$a=679.253(7)$ $c=1098.39(2)$ $a=636.244(9)$ $b=561.653(7)$ $c=607.278(9)$ $\beta=126.1986(8)°$	Not determined
NP 19	$Ni_{45}P_{55}$	900, 15 d	Ni_2P Ni_5P_4 NiP NiP_2	$a=587.60(6)$ $c=338.24(5)$ $a=679.24(2)$ $c=1098.43(5)$ $a=605.06(1)$ $b=488.40(1)$ $c=689.19(2)$ $a=636.90(2)$ $b=561.62(1)$ $c=607.35(2)$ $\beta=126.261(2)°$	Not determined
NP 27	$Ni_{42.5}P_{57.5}$	700, 34 d	Ni_5P_4 NiP_2	$a=679.44(1)$ $c=1098.70(2)$ $a=636.03(2)$ $b=561.99(1)$ $c=607.40(1)$ $\beta=126.162(1)°$	Not determined
NP 27	$Ni_{42.5}P_{57.5}$	880, 3 d	Ni_2P Ni_5P_4 NiP NiP_2	$a=586.06(6)$ $c=339.08(3)$ $a=679.25(2)$ $c=1098.71(6)$ $a=605.06(2)$ $b=488.42(1)$ $c=689.29(2)$ $a=636.90(1)$ $b=561.72(1)$ $c=607.36(1)$ $\beta=126.255(1)°$	Not determined
NP 20	$Ni_{40}P_{60}$	700, 18 d	Ni_5P_4 NiP_2	$a=679.357(9)$ $c=1098.59(2)$ $a=636.580(9)$ $b=561.711(6)$ $c=607.303(8)$ $\beta=126.216(5)°$	Not determined
NP 28	$Ni_{37.5}P_{62.5}$	700, 34 d	Ni_5P_4 NiP_2	$a=679.41(2)$ $c=1098.64(6)$ $a=635.64(2)$ $b=561.74(1)$ $c=607.05(2)$ $\beta=126.155(1)°$	Not determined
NP 32	$Ni_{35}P_{65}$	700, 22 d	Ni_5P_4 NiP_2	$a=679.28(5)$ $c=1098.7(1)$ $a=636.527(7)$ $b=561.668(5)$ $c=607.314(6)$ $\beta=126.2156(5)°$	Not determined
NP 33	$Ni_{33}P_{67}$	200, 27 d	Ni_5P_4 NiP_2 NiP_3	$a=679.41(4)$ $c=1098.5(1)$ $a=634.845(7)$ $b=561.704(5)$ $c=607.256(7)$ $\beta=126.1777(7)°$ $a=782.180(4)$	Not determined
NP 33	$Ni_{33}P_{67}$	400, 17 d	Ni_5P_4 NiP_2 NiP_3	$a=679.38(6)$ $c=1098.9(2)$ $a=635.66(1)$ $b=561.68(1)$ $c=607.14(1)$ $\beta=126.160(1)°$ $a=782.222(7)$	Not determined
NP 33	$Ni_{33}P_{67}$	700, 7 d	Ni_5P_4 NiP_2 NiP_3	$a=679.25(5)$ $c=1098.5(1)$ $a=636.530(8)$ $b=561.636(6)$ $c=607.230(8)$ $\beta=126.2204(7)°$ $a=782.13(1)$	Not determined

No.	Nominal Composition [at %]	Heat Treatment [°C]	Phase	Structure Type	Lattice Param. [pm]	WDS Ni[at.%]	P[at.%]	Σ mass%
NP 34	$Ni_{31}P_{69}$	200, 27 d	Ni_5P_4	Ni_5P_4	a=679.3(1) c=1099.1(4)		Not determined	
			NiP_2	NiP_2	a=635.44(3) b=561.77(2) c=607.25(2) β=126.141(3)°			
			NiP_3	NiP_3	a=782.254(5)			
NP 34	$Ni_{31}P_{69}$	400, 17 d	Ni_5P_4	Ni_5P_4	a=679.6(2) c=1095.5(6)		Not determined	
			NiP_2	NiP_2	a=634.89(4) b=561.68(3) c=606.97(4) β=126.097(4)°			
			NiP_3	NiP_3	a=782.226(1)			
NP 34	$Ni_{31}P_{69}$	700, 7 d	Ni_5P_4	Ni_5P_4	a=679.19(6) c=1098.5(2)		Not determined	
			NiP_2	NiP_2	a=636.39(1) b=561.631(8) c=607.28(1) β=126.211(1)°			
			NiP_3	NiP_3	a=782.105(6)			
NP 38	$Ni_{26}P_{74}$	200, 27 d	NiP_3	NiP_3	a=782.138(4)		Not determined	
			NiP_2	NiP_2	traces only			
NP 38	$Ni_{26}P_{74}$	400, 17 d	NiP_3	NiP_3	a=782.154(4)		Not determined	
NP 38	$Ni_{26}P_{74}$	700, 7 d	NiP_2	NiP_2	a=636.60(2) b=561.64(1) c=607.32(2) β=126.220(2)°		Not determined	
			NiP_3	NiP_3	a=782.113(2)			

d = days

Table 4.2: Experimental results of the thermal analysis in the system Ni–P

No.	Nominal Comp. [at.%]	Heat Treatm. [°C]	DTA heating rate [°C/min]	Heating [°C] Invariant Effects	Heating [°C] Other Effects	Liquidus	Cooling [°C] Liquidus
NP 2	$Ni_{83}P_{17}$	700, 36 d	5	891		904	816
NP 8	$Ni_{81}P_{19}$	700, 36 d	5	892			802
NP 9	$Ni_{80}P_{20}$	700, 36 d	5	889		903	798
NP 3	$Ni_{79}P_{21}$	700, 36 d	5	892		928	834
NP 10	$Ni_{76}P_{24}$	700, 20 d	5	889		965	907
NP 11	$Ni_{74}P_{26}$	700, 36 d	5	976		1028	1004
NP 1	$Ni_{73}P_{27}$	700, 36 d	5	975, 1029		1120	1081
NP 12	$Ni_{72}P_{28}$	700, 36 d	5	1028		1150	1083
alpha	$Ni_{71.6}P_{28.4}$	700, 15d	5	975, 1030		1142	1124
alpha'	$Ni_{71.1}P_{28.9}$	quenched	5	987	1008 max, 1110 eo	1161	1151
NP 4	$Ni_{71}P_{29}$	700, 36 d	5	987	1147 eo	1160	1071
NP 29	$Ni_{70.7}P_{29.3}$	700, 30 d	5	988	1014 max	1161	1136
gamma	$Ni_{70.6}P_{29.4}$	700, 15 d	5	987	1012 max	1161	1154
NP 39	$Ni_{69.5}P_{30.5}$	700, 15 d	5	995	1140 eo	1163	1143
NP 5	$Ni_{69}P_{31}$	700, 36 d	5	994, 1092		1135	1122
NP 13	$Ni_{68}P_{32}$	700, 36 d	5	992, 1090		1127	1111
NP 30	$Ni_{67.2}P_{32.8}$	700, 30 d	5	994, 1095		1108	1089
Ni_2P	$Ni_{66.7}P_{33.3}$	as supplied, powder	5	1096		1105	1081
NP 31	$Ni_{65.5}P_{34.5}$	700, 30 d	5	865		1090	1085
NP 14	$Ni_{63}P_{37}$	700, 36 d	5	867		1058	1047
NP 6	$Ni_{60}P_{40}$	700, 36 d	5	871		1009	1002
NP 21	$Ni_{58}P_{42}$	700, 20 d	5	~870		973	964
NP 22	$Ni_{57}P_{43}$	700, 20 d	5	871		945	915
NP 16	$Ni_{56}P_{44}$	700, 20 d	5	871		917	873
NP 15	$Ni_{55}P_{45}$	700, 36 d	5	861		882	841
NP 15	$Ni_{55}P_{45}$	700, 36 d	2	861		875	837
NP 15	$Ni_{55}P_{45}$	700, 36 d	0.1	860, 863		867	824
NP 23	$Ni_{53.5}P_{46.5}$	700, 34 d	5	859		878	744
NP 23	$Ni_{53.5}P_{46.5}$	700, 34 d	0.1	860, 862		864	832

No.	Nominal Comp. [at.%]	Heat Treatm. [°C]	DTA heating rate [°C/min]	Heating [°C]		Cooling [°C]
				Invariant Effects	Other Effects	Liquidus	Liquidus

No.	Nominal Comp. [at.%]	Heat Treatm. [°C]	DTA heating rate [°C/min]	Invariant Effects	Other Effects	Liquidus	Liquidus
NP 24	$Ni_{52.5}P_{47.5}$	700, 34 d	5	859		878	797
NP 24	$Ni_{52.5}P_{47.5}$	700, 34 d	0.1	861, 864		864	829
NP 25	$Ni_{51.5}P_{48.5}$	700, 34 d	5	859		879	794
NP 25	$Ni_{51.5}P_{48.5}$	700, 34 d	0.1	860, 863		864	836
NP 35	$Ni_{51}P_{49}$	700, 22 d	5	864		881	842
NP 36	$Ni_{49}P_{51}$	700, 22 d	5	861		897	860
NP 26	$Ni_{47.5}P_{52.5}$	700, 30 d	5	860		905	873
NP 19	$Ni_{45}P_{55}$	700, 14 d	5	861		916	887
NP 19	$Ni_{45}P_{55}$	700, 14 d	2	860		908	872
NP 27	$Ni_{42.5}P_{57.5}$	700, 34 d	5	860		922	892
NP 20	$Ni_{40}P_{60}$	700, 18 d	5	859		925	890
NP 20	$Ni_{40}P_{60}$	700, 18 d	2	860		926	894
NP 28	$Ni_{37.5}P_{62.5}$	700, 34 d	5	859		940	---
NP 32	$Ni_{35}P_{65}$	700, 22 d	5	862		940	907

d = days

max = peak maximum

co = extrapolated peak onset

4. Results in the System Ni-P

This Chapter is based on Ref. [89]. Copyright (2009); the full article including Figures and Tables has been reprinted with permission from Elsevier.

The experimental results of the phase analyses based on XRD and EPMA measurements are given in Table 4.1; the maximum stability ranges of the phases derived from these primary data are collected in Table 4.3 together with corresponding information from the literature.

Table 4.3: Maximum stability ranges of binary Ni-P phases

Phase Designation	Structure Type	Composition at. % P		
		This work (experimental)	Oryshchyn et al. [65]	Lee and Nash [58]
(Ni)	Cu	0 - 0.9		0 – 0.32
Ni_3P	Ni_3P	25.0	25.0	25.0
Ni_5P_2 LT	Ni_5P_2	28.4	28.1	28.6
$Ni_{12}P_5$ LT	$Ni_{12}P_5$	29.3	29.3	29.4
Ni_2P	Ni_2P	33.2 – 33.9	33.4 – n.d.	33.3 – n.d.
Ni_5P_4	Ni_5P_4	44.1	-	44.4
NiP	NiP	-	-	50
NiP_2	NiP_2	65.5	-	66.7
NiP_3	NiP_3	-	-	75

n.d. = not determined

Likewise, all results from thermal analyses are listed in Table 4.2, while the derived invariant reactions are collected in Table 4.4.

The phase diagram that was constructed from all these data is shown in Fig. 4.1. It has to be noted that for this particular system the condition of constant pressure, which is usually applied to phase diagrams of intermetallic systems, is not fulfilled over the entire concentration range. This is especially true for the DTA data where the samples are actually heated under their own vapor pressure. In fact, the phase diagram presented here has to be regarded as quasi-isochore[10], because closed quartz crucibles were used for the DTA measurements (see Chapter 3.5). Furthermore, the significant vapor pressure also causes a noticeable change in the sample composition during the

---

[10] Although the volume of the individual quartz crucibles remained practically constant during the DTA measurement, due to the sealing process the different crucibles did not have the same volume. That is why it is preferred to call the system quasi-isochore instead of isochore.

measurement. Therefore a larger uncertainty has to be assumed for the compositions of the DTA data points, especially at higher temperatures[11].

Table 4.4: Invariant Reactions in the System Ni-P according to the present work and the literature.
n.d. = not determined, - = not found to exist

Reaction	Composition of Involved Phases [at.% P]			Temperature [°C], Type and Designation in this work		Literature [58]
				this work		
L = (Ni) + Ni_3P	19.0	0.32	25.0	891, eutectic	e6	870, eutectic
L + Ni_5P_2 LT = Ni_3P	24.5	28.6	25.0	976, peritectic	p7	970, peritectic
L + Ni_5P_2 HT = Ni_5P_2 LT	26.1	28.65	28.6	1029, peritectic	p8	1025, peritectic
Ni_5P_2 HT = Ni_5P_2 LT + $Ni_{12}P_5$ LT	29.0	28.6	29.4	987, eutectoid	e8	1000, eutectoid
$Ni_{12}P_5$ HT = Ni_5P_2 HT + $Ni_{12}P_5$ LT	29.3	29.1	29.4	~ 1005, eutectoid	e10	1025, peritectoid
L + $Ni_{12}P_5$ HT = Ni_5P_2 HT	29.05	29.2	29.4	1161, peritectic	p9	-
L = Ni_5P_2 HT		28.6		-		1170, congruent
L + Ni_5P_2 HT = $Ni_{12}P_5$ HT	30.5	28.6	29.4	-		1125, peritectic
L = $Ni_{12}P_5$ HT		29.4		1170, congruent		-
$Ni_{12}P_5$ HT = $Ni_{12}P_5$ LT + Ni_2P	29.6	29.4	33.2	994, eutectoid	e9	1000, eutectoid
L = $Ni_{12}P_5$ HT + Ni_2P	31.5	30.7	33.2	1092, eutectic	e11	1100, eutectic
L = Ni_2P		33.3		1105, congruent		1110, congruent
L + Ni_2P = Ni_5P_4	45.5	33.9	44.4	870, peritectic	p4	-
L = Ni_2P + Ni_5P_4	40.0	n.d.	44.4	-		880, eutectic
L = Ni_5P_4 + NiP	47.0	44.4	50.0	863, eutectic	e5	-
NiP = Ni_5P_4 + NiP_2	50.0	44.4	66.7	860, eutectoid	e4	850, eutectoid
L + NiP_2 = NiP (tentative)	49.5	66.7	50.0	~890, peritectic	p5	-
Ni_5P_4 + NiP_2 = $Ni_{1.22}P$	44.4	66.7	45.0	-		~ 825, peritectoid
$Ni_{1.22}P$ = Ni_5P_4 + NiP_2	45.0	66.7	44.4	-		~ 770, eutectoid
NiP_3 = NiP_2 + G	66.7	75.0	n.d.	-		700 (estimated)
Metastable						[71]
L = Ni_2P + NiP				n.d.		undetermined, eutectic

---

[11] The error in the composition of the DTA data points is difficult to estimate, as the sample loses P on heating, but (partly) reacts with P from the gas phase again on cooling resulting in a composition shift.

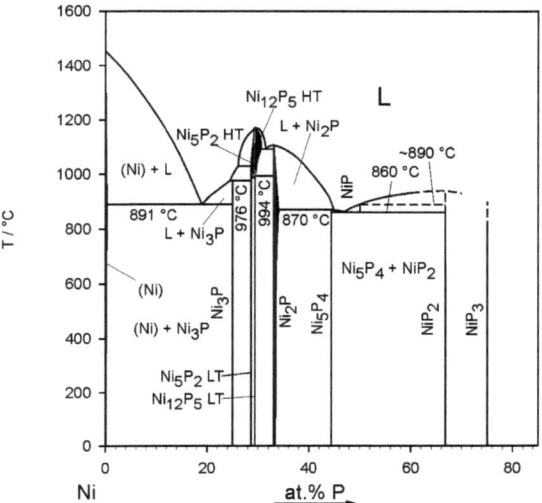

Fig. 4.1: Ni-P Phase Diagram according to the present study; the P-rich part is only tentative.

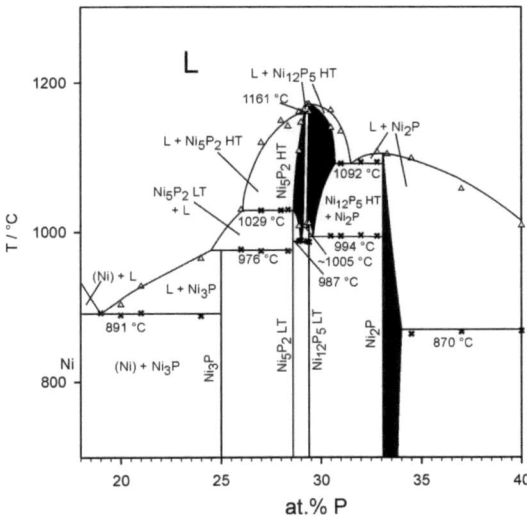

Fig. 4.2: Ni-rich section of the Ni-P Phase Diagram with data points from DTA: x, invariant effects; Δ, other effects.

4.1 The Ni-rich section between 0 and 35 at.% P

Fig. 4.2 shows an enlarged section of the Ni-rich part of the Ni-P phase diagram together with experimental data points from the DTA measurements. In this composition range, the phases Ni_3P, Ni_5P_2 LT, $Ni_{12}P_5$ LT and Ni_2P were found as reported in literature [58, 65]. The phase compositions determined by EPMA in the present study are in good agreement with the literature results.

All phases, except for Ni_2P, were found to be line compounds, and the P contents of 28.4 and 29.3 at.% P, as determined by EPMA for Ni_5P_2 LT and $Ni_{12}P_5$ LT, agree nicely with the stoichiometric values of 28.6 and 29.4 at.% P. Therefore, the two phases are shown at their respective stoichiometric compositions in the phase diagram. In the literature a significant homogeneity range of Ni_2P was assumed to exist and was shown as such in previous phase diagram versions. According to the EPMA data in the present work there is indeed a small homogeneity range of Ni_2P from 33.2 to 33.8 at.% P at 700 °C. The P-rich phase boundary was investigated in more detail between 550 and 810 °C, and a slight widening of the homogeneity range could be observed, i.e., the P-rich stability limit was found to increase from 33.5 at.% P at 550 °C to 33.9 at.% at 810 °C (see Table 4.1).

In case of Ni_5P_2 LT difficulties were encountered when trying to index the experimental XRD pattern of this phase using the structural model from Oryshchyn et al. [65], and only a qualitatively satisfying description could be achieved. Further single crystal studies of this phase are planned in order to clarify the remaining uncertainties.

For the phases Ni_5P_2 and $Ni_{12}P_5$ the existence of HT modifications was concluded from thermal analysis in the available literature, but the crystal structures of these phases have not been determined so far. Therefore several attempts were made to quench samples of these phases. However, in practically all samples annealed at temperatures above the reported transition temperatures, quenched either by immersing the quartz capsule into water or by even breaking the quartz glass tube on contact with the water, the diffraction patterns of the corresponding LT modifications were found in the XRD measurements. Nevertheless, in the DTA curves several thermal effects were found in this area at 1029, 987 and 994 °C. Thus it is concluded that the HT modifications cannot be retained by quenching and that the corresponding transition reactions take place at very high reaction rates, a phenomenon not uncommon with intermetallic HT phases.

The observed thermal effects were assigned to the following invariant reactions: the peritectic p8, L + Ni_5P_2 HT = Ni_5P_2 LT (1029 °C), the eutectoid e8, Ni_5P_2 HT = Ni_5P_2 LT + $Ni_{12}P_5$ LT (987 °C), and the eutectoid e9, $Ni_{12}P_5$ HT = $Ni_{12}P_5$ LT + Ni_2P (994 °C). A further invariant reaction was included in the phase diagram as required by the above mentioned reaction sequence, i.e. the eutectoid reaction e10, $Ni_{12}P_5$ HT = Ni_5P_2 HT + $Ni_{12}P_5$ LT. As this reaction was not unequivocally observed in any of the DTA recordings, it was tentatively placed at approx. 1005 °C. In this way, $Ni_{12}P_5$ LT transforms congruently into its HT modification.

In general, the region of the two HT phases appears to be quite different than reported in the literature. For both HT phases a homogeneity range was introduced which, however, could not be retained during quenching of relevant samples. This phase diagram version is therefore exclusively based on a high number of monovariant effects in the DTA measurements (e.g. in sample NP 29 at 29.5 at.% P). For the interpretation of these monovariant effects the onset temperature of the peak on heating was taken for the precipitation of an additional phase (i.e. the beginning of a two-phase field) and the maximum was taken for the end of this precipitation (i.e. the end of a two-phase field). This situation is shown in Fig. 4.2 together with the data points from DTA.

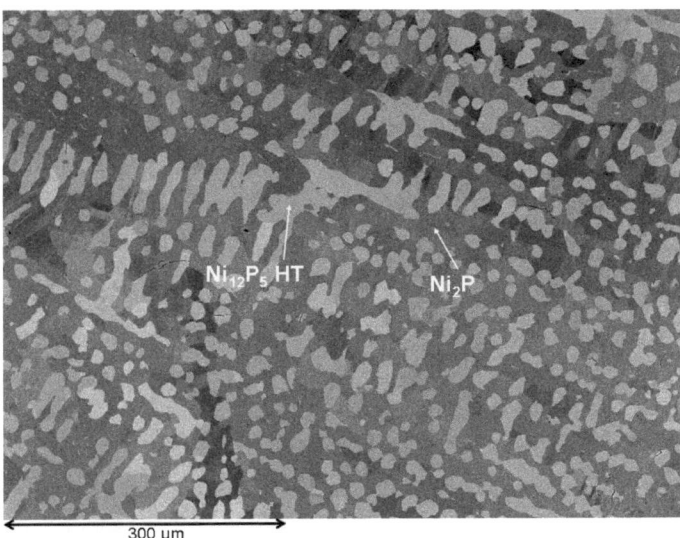

Fig. 4.3: SEM image of sample NP 13 ($Ni_{68}P_{32}$) cooled down from the melt on air. The microstructure resulting from the eutectic e11, L = $Ni_{12}P_5$ HT + Ni_2P can be seen. Different shades in the light and dark phases are due to different orientation of the individual grains.

In the literature (Yupko et al. [64]) the peritectic formation of $Ni_{12}P_5$ HT through L + Ni_5P_2 HT = $Ni_{12}P_5$ LT had been reported at about 1120 °C. None of the DTA recordings carried out for the present work showed any thermal effect at this temperature. Instead, invariant effects were observed at 1161 °C (e.g. in sample NP 29). Together with the observed liquidus effects in this composition range, this rather suggests the peritectic formation of Ni_5P_2 HT out of $Ni_{12}P_5$ HT in a reaction p9, L + $Ni_{12}P_5$ HT = Ni_5P_2 HT, which is the opposite of the literature version.

Thus, according to the present data, $Ni_{12}P_5$ HT plays an important role for the solidification behavior of alloys in this area: it is formed congruently out of the melt and transforms itself congruently into the LT modification (comprising two eutectoid reactions). The other phases Ni_5P_2 HT, Ni_5P_2 LT and Ni_3P are formed by a cascade of the peritectic reactions p9, p8 and p7 (at 1161, 1029 and 976 °C, respectively, see also Table 4.4) starting from $Ni_{12}P_5$ HT, too.

The temperatures of all other invariant reactions were checked by DTA. While the obtained phase equilibria are in principle in agreement with existing literature reports, the present reaction temperatures deviate to some extent from the reported ones: for example, a temperature of 891 °C was obtained for the eutectic e6, L = (Ni) + Ni_3P instead of 870 °C [58] (see also Table 4.4). Effects observed at 1092 °C correspond to the eutectic reaction e11, L = $Ni_{12}P_5$ HT + Ni_2P, and this result is consistent with the literature reports: Fig. 4.3 shows the typical eutectic microstructure obtained in sample NP 13 ($Ni_{68}P_{32}$) cooled down from the melt by removing the quartz capsule from the furnace and cooling it on air.

4.2 The central part between 35 and 66.7 at.% P

Fig. 4.4 shows the enlarged central part of the phase diagram based on the present results. It should be pointed out that at these compositions the phase diagram is clearly not isobaric due to high P vapor pressures.

Ni_5P_4 was placed at its nominal P content of 44.4 at.% P because of its good agreement with the determined EPMA value of 44.1 at.% P. It was found to be formed by a peritectic reaction, i.e. p4, L + Ni_2P = Ni_5P_4. The temperature of this reaction was determined to be 870 °C according to DTA results. This temperature does not agree with literature data in this region, where a eutectic reaction is reported at 830 °C [61] or 880 °C [58]. However, the present interpretation is supported by a metallographic investigation of samples NP 16, 21 and 22 within the two phase region [L + Ni_2P], annealed and quenched from 900 °C, where primary crystallization of Ni_2P surrounded by a rim of Ni_5P_4 and a eutectic matrix was found (see also Fig. 4.5).

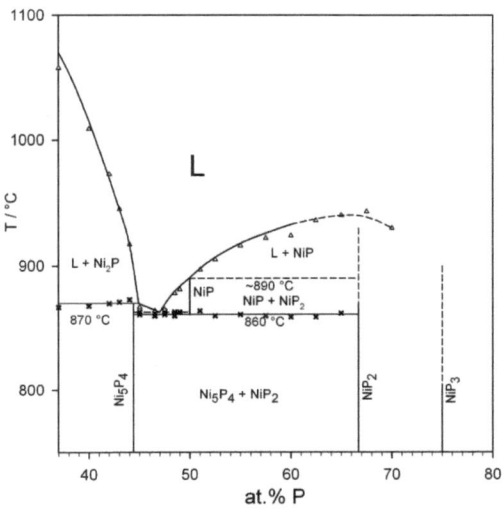

Fig. 4.4: Central part of the Ni-P Phase Diagram with data points from DTA as in Fig. 4.2.

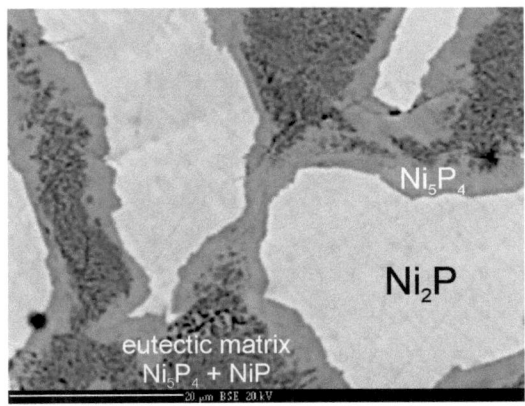

Fig. 4.5: SEM image of sample NP 21 ($Ni_{58}P_{42}$) quenched from 900 °C showing the primary crystallization of Ni_2P, the peritectic rim of Ni_5P_4 and the eutectic matrix (Ni_5P_4 + NiP according to XRD).

Unfortunately, the matrix was too fine to allow the determination of its constituents by EPMA, but XRD clearly showed the presence of NiP which has to be one of the constituents of this eutectic matrix. This result is also in contrast to the evaluation of Larsson [71], who proposed the peritectic reaction L + NiP = Ni$_5$P$_4$. However, this reaction had only been based on the optical appearance and shape of the sample pellets after water quenching, i.e. whether they had changed shape due to melting in the silica tube or retained an irregular shape (for details see [71]). No thermal analyses or metallographic investigations had been performed by Larsson.

The phase NiP itself was reported by Larsson [71]. In the present work this phase was found in the XRD analyses of samples annealed at 900 °C (for the problems related to these samples see below) together with the surrounding phases (see Table 4.1). In the literature NiP is claimed to be stable above approximately 850 °C. Indeed, the corresponding DTA curves in the present work showed an invariant effect at 860 °C which was assigned to the eutectoid decomposition e4, NiP = Ni$_5$P$_4$ + NiP$_2$. At all temperatures below this reaction, the two phase field [Ni$_5$P$_4$ + NiP$_2$] was observed in the XRD measurements of corresponding samples. An additional phase Ni$_{1.22}$P which had been reported in the work of Larsson [71] was neither found by XRD nor EPMA, as all reflexes in the powder patterns could be indexed perfectly well and none remained unaccounted for. No crystallographic information for this phase is available in the literature, and it was reported by no other author either (although Lee and Nash [58] claim that Ni$_{1.22}$P was confirmed by Yupko et al. [64], this is incorrect and rather the result of a misunderstanding of the text in Ref. [64]). Furthermore, no thermal effects could be detected in the present study, neither in the first nor in the second heating cycle, that would correspond to the formation or decomposition of this phase, respectively. Therefore, this phase was not included in the present version of the phase diagram. However, only high temperature XRD would allow a definite clarification of this situation but this is hampered by the high vapor pressure and loss of P at these concentrations.

At temperatures above the eutectoid reaction at 860 °C, the interpretation of the experimental results is complicated by the considerable evaporation of P. All samples annealed in this range were hollow, i.e. they consisted of a rather thin and brittle metal layer on the inner quartz surface left after annealing. Though XRD of this material was possible, it could not be embedded for EPMA investigations. Furthermore, XRD frequently revealed the presence of up to four phases, e.g. in samples NP 19 and 27 (Ni$_{45}$P$_{55}$ and Ni$_{42.5}$P$_{57.5}$, respectively) quenched from 880 °C or higher. While the existence of NiP could thus be proven, the presence of Ni$_5$P$_4$ and Ni$_2$P was due to non-equilibrium. In the literature [71] a metastable eutectic L = Ni$_2$P + NiP had been reported, where the formation of Ni$_5$P$_4$ is completely suppressed. The appearance of Ni$_2$P in samples NP 19 and 27 further supports the existence of such a metastable reaction.

Between the eutecitoid reaction e4, NiP = Ni_5P_4 + NiP_2 at 860 °C and the peritectic p4, L + Ni_2P = Ni_5P_4 at 870 °C, a eutectic reaction L = Ni_5P_4 + NiP (e5) is to be expected in order not to violate Gibbs' phase rule, and in fact a eutectic matrix can be seen in the micrograph in Fig. 4.5. In the DTA curves at 5 K/min only one thermal effect was observed. In order to resolve these apparently very close effects in the DTA curves, additional measurements were carried out at heating rates of 2 and 0.1 K/min. Indeed, the latter ones revealed the existence of several peaks that could not be resolved at the standard heating rate of 5 K/min. As an example, the relevant section of the DTA curve of sample NP 15, $Ni_{55}P_{45}$, is shown in Fig. 4.6. From these measurements the temperature of 863 °C was deduced for e5, L = Ni_5P_4 + NiP, together with an accurate value for the liquidus temperature, despite a rather high background noise.

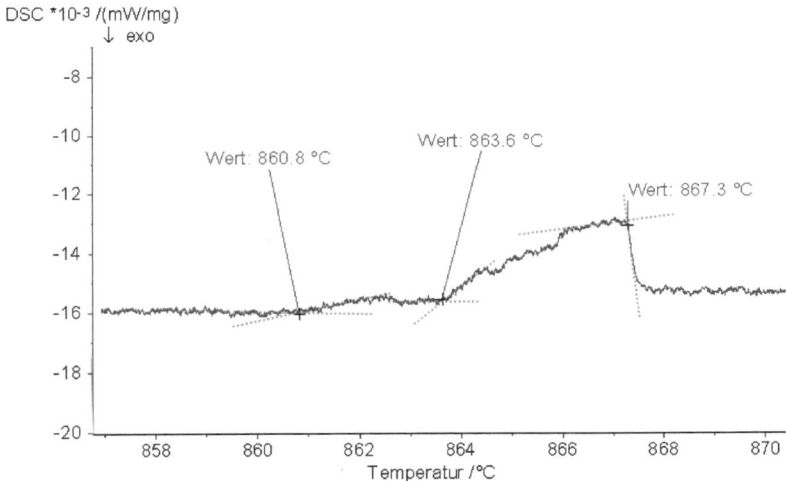

Fig. 4.6: Section of the DTA heating curve of sample NP 15 ($Ni_{55}P_{45}$) using a heating rate of 0.1 K/min: three thermal effects can be distinguished as explained in the text.

On the P-rich side of NiP a further invariant reaction has to exist corresponding to its melting. In the DTA curves of some of the samples in the corresponding composition range various thermal effects between 890 °C and 920 °C were obtained that may indicate this reaction. However, no clear interpretation of these effects was possible due to the large scatter and overlap with other peaks.

The liquidus, as derived from our DTA recordings, was found to rise from the eutectic L = Ni_5P_4 + NiP almost continually towards more P-rich compositions. Its temperature is in agreement with the observation of Biltz and Heimbrecht [73] who reported the melting of remaining material during their degradation experiments at approximately 900 °C in the relevant composition region. According to the course of the liquidus the invariant reaction at about 890 °C can be a peritectic (p5, L + NiP$_2$ = NiP) as indicated by a dashed line in Figs. 4.1 and 4.4 or a very shallow eutectic (L = NiP + NiP$_2$).

4.3 The P-rich section with P contents of more than 66.7 at.%

Three samples were prepared in this section (samples NP 33, 34 and 38, cf. Table 4.1) and annealed at 200, 400 and 700 °C. These samples look clearly non-metallic, as can be seen in Fig. 4.7. As they were extremely fragile, it was not possible to grind and polish them for use in the SEM, and no EPMA values could be produced. According to XRD analyses, samples NP 33 and 34 contained the three phases NiP$_3$, NiP$_2$ and Ni$_5$P$_4$ (traces), which may be caused by the influence of the gas phase.

Fig. 4.7: Ni-P samples NP 33, 34 and 38 with 67, 69 and 74 at.% P (from left to right); these samples were prepared from powders of Ni and red P, pressed into pellets and annealed at 700 °C: their clearly non-metallic appearance is evident.

In contrast to the literature, where a lower stability limit of 700 °C for NiP$_3$ and a eutectoid decomposition were reported [58], XRD results showed that NiP$_3$ exists at 200 and 400 °C as well. This is supported by results from the DTA, where certainly no thermal effect around 700 °C was observed, although the DTA measurements in this region produced some confusing results above 700 °C. DTA measurements in this area were finally discontinued after attempts to minimize the loss of P by reduction of the crucible volume and filling with 0.5 bar Ar before sealing were recognized as unsuccessful, and after a rather destructive explosion of a sample during the measurement had been encountered. A rather rudimentary representation of this section is included in Fig. 4.4, where the stability range of NiP$_3$ has been extended down to room temperature but no information can be given on its melting behavior. No samples with P-contents higher than 75 at.% were investigated in the present work. Therefore no phase diagram information can be given beyond this P-amount.

Table 5.1: Experimental results of the phase analysis in the system Ni-P-Sn

No.	Nominal Composition [at.%]	Heat Treatment [°C]	Phase Analysis			EDX / WDS [at.%]			Σ wt.%
			Phase	Structure Type	Lattice Param. [pm]	Ni	P	Sn	
850 °C									
NPS 3	$Ni_{63.33}P_{31.67}Sn_5$	850, 37d	Sn (I)	βSn	a=583.39(2) c=318.30(1)	not determined			
			Ni_2P	Fe_2P	a=586.94(2) c=338.42(2)				
NPS 4	$Ni_{60}P_{30}Sn_{10}$	850, 37d	Sn (I)	βSn	a=583.05(3) c=318.11(2)	3.1	0.0	96.9	101.5
			Ni_3Sn_2 HT	$InNi_2$	not found in XRD	50.4	19.5	30.1	101.1
			Ni_3Sn_4	Ni_3Sn_4	not found in XRD	41.8	0.0	58.2	100.9
			Ni_2P	Fe_2P	a=585.65(3) c=338.81(2)	65.7	34.3	0.0	98.9
NPS 6	$Ni_{53.33}P_{26.67}Sn_{20}$	850, 37d	Sn (I)	βSn	a=583.141(8) c=318.288(5)	0.7	0.0	99.3	101.1
			Ni_3Sn_2 HT	$InNi_2$	a=365.34(2) c=512.20(5)	not found in EPMA			
			Ni_2P	Fe_2P	a=586.41(2) c=338.58(2)	65.5	34.5	0.0	98.3
NPS 19	$Ni_{61.7}P_{8.3}Sn_{30}$	850, 59d	Ni_3Sn_2 HT	$InNi_2$	a=403.902(6) c=516.73(1)	59.2	4.6	36.2	101.1
			$Ni_{12}P_5$ LT	$Ni_{12}P_5$	a=864.58(3) c=506.88(2)	70.1	29.7	0.2	99.1
NPS 20	$Ni_{63}P_{17}Sn_{20}$	850, 59d	Ni_3Sn_2 HT	$InNi_2$	a=397.04(1) c=518.242(2)	58.9	9.6	31.5	101.0
			$Ni_{12}P_5$ LT	$Ni_{12}P_5$	a=864.96(3) c=507.42(3)	70.1	29.7	0.2	99.1
NPS 21	$Ni_{65}P_{25}Sn_{10}$	850, 59d	Ni_3Sn_2 HT	$InNi_2$	a=390.96(1) c=519.82(2)	58.5	12.6	28.9	101.5
			$Ni_{12}P_5$ LT	$Ni_{12}P_5$	a=864.39(3) c=507.17(2)	70.2	29.7	0.1	99.6
			Ni_2P	Fe_2P	a=586.35(4) c=339.21(3)	66.4	33.4	0.2	99.6
NPS 22	$Ni_{72.92}P_{8.33}Sn_{18.75}$	850, 16d	Ni_3Sn HT	BiF_3	a=585.44(1)	74.4	2.6	23.0	101.2
			Ni_3Sn_2 HT	$InNi_2$	a=413.95(1) c=521.00(2)	63.1	0.9	36.0	101.7
			T2	C_6Cr_{23}	a=1111.58(2)	72.1	20.8	7.1	100.3
NPS 23	$Ni_{70.8}P_{16.7}Sn_{12.5}$	850, 16d	Ni_3Sn HT	BiF_3	not found in XRD	74.4	2.6	23.1	101.4
			Ni_3Sn_2 HT	$InNi_2$	a=414.115(7) c=521.19(1)	63.0	0.9	36.1	100.9
			T2	C_6Cr_{23}	a=1111.74(1)	71.9	20.9	7.2	100.5
NPS 24	$Ni_{68.75}P_{25}Sn_{6.25}$	850, 16d	Ni_3Sn_2 HT	$InNi_2$	a=406.493(5) c=516.638(9)	60.4	2.7	36.9	102.2
			$Ni_{12}P_5$ LT	$Ni_{12}P_5$	a=864.627(7) c=507.191(5)	70.2	29.7	0.1	99.8
NPS 25	$Ni_{71.25}P_5Sn_{23.75}$	850, 19d	Ni_3Sn HT	BiF_3	a=585.57(1)	61.1	1.5	37.4	100.2
			Ni_3Sn_2 HT	$InNi_2$	a=413.971(7) c=521.03(1)	71.0	21.5	7.5	99.6
			T2	C_6Cr_{23}	a=1111.49(3)	73.1	2.7	26.5	100.8

No.	Nominal Composition [at.%]	Heat Treatment [°C]	Phase Analysis					
			Phase	Structure Type	Lattice Param. [pm]	EDX / WDS [at.%]		Σ wt.%
						Ni	P	Sn

No.	Nominal Composition [at.%]	Heat Treatment [°C]	Phase	Structure Type	Lattice Param. [pm]	Ni	P	Sn	Σ wt.%
850 °C									
NPS 26	Ni$_{67.5}$P$_{10}$Sn$_{22.5}$	850, 19d	Ni$_3$Sn$_2$ HT	InNi$_2$	a=414.02(1) c=521.11(2)	not determined			
			T2	C$_6$Cr$_{23}$	a=1111.36(4)				
NPS 27	Ni$_{63.75}$P$_{15}$Sn$_{21.25}$	850, 19d	Ni$_3$Sn$_2$ HT	InNi$_2$	a=403.61(2) c=516.37(3)	57.8	4.8	37.4	100.3
			Ni$_{12}$P$_5$ LT	Ni$_{12}$P$_5$	a=864.15(6) c=507.06(4)	69.6	30.3	0.2	96.7
NPS 28	Ni$_{60}$P$_{20}$Sn$_{20}$	850, 19d	Ni$_3$Sn$_2$ HT	InNi$_2$	a=387.16(3) c=521.10(4)	56.8	14.5	28.7	98.1
			Ni$_2$P	Fe$_2$P	a=586.34(8) c=339.18(6)	65.7	34.1	0.2	95.8
NPS 29[a)]	Ni$_{56.25}$P$_{25}$Sn$_{18.75}$	850, 19d	Sn (l)	βSn	a=583.32(1) c=318.29(1)	8.3	2.6	89.1	99.0
			Ni$_3$Sn$_2$ HT	InNi$_2$	a=377.33(2) c=520.59(5)	53.5	17.0	29.5	100.5
			Ni$_3$Sn$_4$ (l)	Ni$_3$Sn$_4$	traces only	42.8	0.0	57.2	101.9
			Ni$_2$P	Fe$_2$P	a=586.50(3) c=338.53(2)	65.7	34.2	0.2	98.4
			T5 (l)	Ni$_2$PSn	a=1271(2) b=367.6(5) c=526.7(4)	not found in EPMA			
NPS 30	Ni$_{52.5}$P$_{30}$Sn$_{17.5}$	850, 19d	Sn (l)	βSn	a=583.497(9) c=318.464(6)	2.5	0.0	97.5	100.3
			Ni$_2$P	Fe$_2$P	a=585.93(6) c=338.81(6)	65.5	34.4	0.1	97.3
			Ni$_5$P$_4$	Ni$_5$P$_4$	a=680.1(1) c=1100.6(4)	not found in EPMA			
			T5 (l)	Ni$_2$PSn	a=1260.7(5) 359.6(2) c=511.8(2)	not found in EPMA			
NPS 34	Ni$_{54}$P$_{10}$Sn$_{36}$	850, 19d	Sn (l)	βSn	a=583.4(2) c=326.8(2)	0.8	0.0	99.2	101.0
			Ni$_3$Sn$_2$ HT	InNi$_2$	a=390.238(7) c=519.54(1)	53.1	12.6	34.3	100.0
			Ni$_3$Sn$_4$ (l)	Ni$_3$Sn$_4$	traces only	45.0	0.0	55.0	99.7
NPS 35	Ni$_{51}$P$_{15}$Sn$_{34}$	850, 19d	Sn (l)	βSn	a=583.1(2) c=318.3(2)	1.2	0.0	98.8	101.2
			Ni$_3$Sn$_2$ HT	InNi$_2$	a=381.66(1) c=521.95(2)	54.3	16.2	29.5	100.6
			Ni$_3$Sn$_4$ (l)	Ni$_3$Sn$_4$	traces only	42.9	0.0	57.1	99.8
NPS 36	Ni$_{48}$P$_{20}$Sn$_{32}$	850, 19d	Sn (l)	βSn	a=583.278(7) c=318.257(5)	1.3	0.0	98.7	101.4
			Ni$_3$Sn$_2$ HT	InNi$_2$	a=379.75(2) c=520.58(4)	52.3	18.1	29.6	99.3
			Ni$_3$Sn$_4$ (l)	Ni$_3$Sn$_4$	a=1223.6(2) b=405.19(6) c=522.61(7) β=105.18(1)°	43.5	0.0	56.5	99.6
			Ni$_2$P	Fe$_2$P	a=586.29(3) c=339.20(2)	64.8	34.2	1.0	96.7
			T3	Ni$_{10}$P$_3$Sn$_5$	a=644.3(1) b=838.3(2) c=1006.0(1) α=75.25(2)° β=83.46(2)° γ=84.14(1)°	52.3	15.4	32.4	98.9
NPS 37	Ni$_{45}$P$_{25}$Sn$_{30}$	850, 19d	Sn (l)	βSn	a=583.176(3) c=318.260(3)	0.5	0.0	99.5	99.9
			Ni$_2$P	Fe$_2$P	a=586.44(6) c=338.16(5)	64.9	35.0	0.2	98.8
			P$_3$Sn$_4$ (l)	Bi$_3$Se$_4$	a=396.68(7) c=3530(1)	0.0	38.9	61.2	102.4
			T5 (l)	Ni$_2$PSn	a=1269.3(3) b=359.30(7) c=509.8(1)	25.9	52.5	21.6	102.2

Sample	Composition	T(°C), time	Phases	Structure type	Lattice parameters	Ni	P	Sn	Total
NPS 38	Ni$_{42}$P$_{30}$Sn$_{28}$	850, 19d	Sn (l)	βSn	no XRD made	1.9	0.0	98.1	102.4
			Ni$_2$P	Fe$_2$P		65.2	34.8	0.0	97.6
			Ni$_5$P$_4$	Ni$_5$P$_4$		54.2	45.3	0.5	97.7
			NiP$_2$	NiP$_2$		32.5	67.3	0.2	98.1
			P$_3$Sn$_4$ (l)	Bi$_3$Se$_4$		1.5	41.1	57.4	102.1
NPS 50	Ni$_{71.9}$P$_{12.5}$Sn$_{15.6}$	850, 19d	Ni$_3$Sn HT	BiF$_3$	a=585.30(2)	74.7	2.5	22.8	101.2
			Ni$_3$Sn$_2$ HT	InNi$_2$	a=413.88(1) c=520.65(2)	63.5	0.9	35.6	101.5
			T2	C$_6$Cr$_{23}$	a=1111.03(2)	72.3	20.6	7.2	100.4
NPS 51	Ni$_{69.8}$P$_{20.8}$Sn$_{9.4}$	850, 19d	Ni$_3$Sn$_2$ HT	InNi$_2$	a=406.420(7) c=516.34(1)	60.4	2.5	37.1	102.0
			Ni$_{12}$P$_5$ LT	Ni$_{12}$P$_5$	a=864.60(2) c=507.12(2)	70.2	29.6	0.2	99.8
			T1	Ni$_{10}$P$_3$Sn	a=766.497(8) c=962.14(1)	71.2	21.7	7.1	100.6
NPS 52	Ni$_{67.7}$P$_{29.2}$Sn$_{3.1}$	850, 19d	Ni$_3$Sn$_2$ HT	InNi$_2$	a=391.06(1) c=519.57(2)	58.3	12.4	29.3	102.2
			Ni$_{12}$P$_5$ LT	Ni$_{12}$P$_5$	a=864.21(2) c=507.03(2)	70.4	29.6	0.0	100.2
			Ni$_2$P	Fe$_2$P	a=586.54(2) c=338.86(2)	66.6	33.3	0.1	100.7
NPS 53	Ni$_{60.8}$P$_{4.2}$Sn$_{35}$	850, 19d	Ni$_3$Sn$_2$ HT	InNi$_2$	a=407.846(4) c=517.274(5)	60.2	2.3	37.5	101.3
			Ni$_{12}$P$_5$ LT	Ni$_{12}$P$_5$	a=864.6(2) c=507.4(1) (traces)	70.1	29.8	0.1	99.5
			T1	Ni$_{10}$P$_3$Sn		71.2	21.7	7.1	100.4
NPS 54	Ni$_{62.5}$P$_{12.5}$Sn$_{25}$	850, 19d	Ni$_3$Sn$_2$ HT	InNi$_2$	a=403.696(5) c=516.892(7)	59.4	4.7	35.9	100.5
			Ni$_{12}$P$_5$ LT	Ni$_{12}$P$_5$	a=864.79(2) c=507.14(2)	70.1	29.7	0.2	99.7
NPS 55	Ni$_{64.2}$P$_{20.8}$Sn$_{15}$	850, 19d	Ni$_3$Sn$_2$ HT	InNi$_2$	a=392.053(7) c=519.31(1)	58.8	12.1	29.1	101.8
			Ni$_{12}$P$_5$ LT	Ni$_{12}$P$_5$	a=864.33(2) c=507.10(1)	70.3	29.5	0.2	99.7
NPS 56	Ni$_{65.8}$P$_{29.2}$Sn$_5$	850, 19d	Ni$_3$Sn$_2$ HT	Fe$_2$P	a=391.143(9) c=519.85(2)	58.5	12.5	29.0	102.1
			Ni$_{12}$P$_5$ LT	Ni$_{12}$P$_5$	a=864.52(2) c=507.29(1)	not found in EPMA			
			Ni$_2$P	Fe$_2$P	a=586.82(1) c=339.01(1)	66.6	33.3	0.1	100.3
NPS 62	Ni$_{74.2}$P$_{23}$Sn$_{2.5}$	550, 59d	(Ni)	Cu	a=354.33(2)	not determined			
			Ni$_3$P	Ni$_3$P	a=865.520(5) c=438.83(3)				
			T2	C$_6$Cr$_{23}$	a=1111.626(7)				
NPS 63	Ni$_{90}$P$_5$Sn$_5$	700, 19d	(Ni)	Cu	a=357.860(6)	not determined			
			T2	C$_6$Cr$_{23}$	a=1110.95(2)				
NPS 64	Ni$_{80}$P$_5$Sn$_{15}$	700, 19d	(Ni)	Cu	a=358.46(3)	not determined			
			Ni$_3$Sn HT	Mg$_3$Cd	a=585.58(5)				
			T2	C$_6$Cr$_{23}$	a=1111.7(1)				
NPS 65	Ni$_{90}$P$_{15}$Sn$_5$	700, 19d	(Ni)	Cu	a=354.35(1)	not found in EPMA			
			Ni$_3$P	Ni$_3$P	a=895.63(7) c=438.88(5)				
			T2	C$_6$Cr$_{23}$	a=1111.65(4)				
NPS 79	Ni$_{71}$P$_{27}$Sn$_2$	850, 24d	Ni$_5$P$_2$ LT	Ni$_5$P$_2$	a=661.59(6) c=1230.2(2)	not found in EPMA			
			Ni$_{12}$P$_5$ LT	Ni$_{12}$P$_5$	a=864.964(8) c=507.027(6)	70.3	29.6	0.1	100.4
			T1	Ni$_{10}$P$_3$Sn	a=767.48(1) c=962.46(2)	71.4	21.5	7.1	100.6

No.	Nominal Composition [at.%]	Heat Treatment [°C]	Phase Analysis					EDX / WDS [at.%]			Σ wt.%
			Phase	Structure Type	Lattice Param. [pm]	Ni	P	Sn	
850 °C									
NPS 79b	$Ni_{73}P_{25}Sn_{2}$	850, 24d	Ni_3P T2 T	Ni_3P C_6Cr_{23} ---	$a=889.27(7)$ $c=450.12(7)$ $a=1111.09(4)$ ---	75.1 72.3 72.9	24.9 20.4 26.3	0.0 7.3 0.8	100.1 100.3 99.9
NPS 104	$Ni_{70.5}P_{18}Sn_{11.5}$	850, 6d	Ni_3Sn_2 HT T1 T2	$InNi_2$ $Ni_{10}P_3Sn$ C_6Cr_{23}	$a=412.118(8)$ $c=519.76(2)$ $a=768.31(1)$ $c=961.98(2)$ $a=1111.62(2)$	not determined			
NPS 105	$Ni_{70.4}P_{18.5}Sn_{11.1}$	850, 10d	Ni_3Sn_2 HT T1 T2	$InNi_2$ $Ni_{10}P_3Sn$ C_6Cr_{23}	$a=411.993(6)$ $c=519.65(1)$ $a=768.250(7)$ $c=961.96(1)$ $a=1111.72(4)$	not determined			
NPS 106	$Ni_{70.3}P_{19}Sn_{10.7}$	850, 10d	Ni_3Sn_2 HT T1	$InNi_2$ $Ni_{10}P_3Sn$	$a=410.504(6)$ $c=518.54(1)$ $a=768.054(6)$ $c=962.09(1)$	not determined			
NPS 107	$Ni_{69.5}P_{22}Sn_{8.5}$	850, 10d	Ni_3Sn_2 HT $Ni_{12}P_5$ LT T1	$InNi_2$ $Ni_{12}P_5$ $Ni_{10}P_3Sn$	$a=406.940(7)$ $c=516.30(1)$ $a=864.89(1)$ $c=507.186(9)$ $a=766.980(7)$ $c=962.18(1)$	not determined			
NPS 108	$Ni_{69.25}P_{23}Sn_{7.75}$	850, 10d	Ni_3Sn_2 HT $Ni_{12}P_5$ LT T1	$InNi_2$ $Ni_{12}P_5$ $Ni_{10}P_3Sn$	$a=407.108(7)$ $c=516.41(1)$ $a=864.64(1)$ $c=507.159(8)$ $a=767.168(9)$ $c=962.13(2)$	not determined			
NPS 109	$Ni_{69}P_{28}Sn_{4}$	850, 6d	Ni_3Sn_2 HT $Ni_{12}P_5$ LT	$InNi_2$ $Ni_{12}P_5$	$a=407.394(5)$ $c=516.79(1)$ $a=864.698(8)$ $c=507.209(6)$ + unindexed lines	not determined			
NPS 110	$Ni_{67.3}P_{31}Sn_{1.7}$	850, 10d	Ni_3Sn_2 HT $Ni_{12}P_5$ LT Ni_2P	$InNi_2$ $Ni_{12}P_5$ Ni_2P	$a=391.83(1)$ $c=519.19(4)$ $a=864.66(1)$ $c=507.193(7)$ $a=586.49(3)$ $c=339.09(3)$	not determined			
NPS 111	$Ni_{67.3}P_{31}Sn_{1.7}$	850, 5d	Ni_3Sn_2 HT $Ni_{12}P_5$ LT Ni_2P	$InNi_2$ $Ni_{12}P_5$ Ni_2P	$a=391.21(2)$ $c=519.35(4)$ $a=864.67(1)$ $c=507.166(8)$ $a=586.48(3)$ $c=339.03(3)$	not determined			
NPS 112	$Ni_{65.5}P_{27}Sn_{7.5}$	850, 5d	Ni_3Sn_2 HT $Ni_{12}P_5$ LT Ni_2P	$InNi_2$ $Ni_{12}P_5$ Ni_2P	$a=391.514(9)$ $c=519.59(2)$ $a=864.65(2)$ $c=507.23(1)$ $a=586.56(3)$ $c=339.06(3)$	not determined			
NPS 113	$Ni_{64.7}P_{23}Sn_{12.3}$	850, 4d	Ni_3Sn_2 HT $Ni_{12}P_5$ LT Ni_2P	$InNi_2$ $Ni_{12}P_5$ Ni_2P	$a=391.769(5)$ $c=519.48(1)$ $a=864.63(1)$ $c=507.236(8)$ $a=586.53(4)$ $c=339.14(3)$	not determined			

Sample	Composition	T, t	Phases	Lattice parameters (pm)			
NPS 114	Ni$_{63.7}$P$_{18}$Sn$_{18.3}$	850, 5d	InNi$_2$ Ni$_{12}$P$_5$	a=397.296(5) c=517.84(1) a=864.66(1) c=507.26(1)	not determined		
NPS 115	Ni$_{62.8}$P$_{14}$Sn$_{23.2}$	850, 4d	Ni$_3$Sn$_2$ HT Ni$_{12}$P$_5$ LT	a=400.523(7) c=516.79(2) a=864.50(2) c=507.13(2)	not determined		
NPS 116	Ni$_{72}$P$_{26}$Sn$_2$	850, 8d	Ni$_5$P$_2$ LT Ni$_{10}$P$_3$Sn T1 C$_6$Cr$_{23}$ T2	a=661.705(7) c=1232.94(2) a=767.03(2) c=963.50(5) a=1111.62(2)	not determined		
NPS 117	Ni$_{74}$P$_{24}$Sn$_2$	850, 8d	Ni$_3$P T2 C$_6$Cr$_{23}$ T	a=895.60(1) c=438.837(9) a=1111.74(2) ---	not determined		
NPS 118	Ni$_{71}$P$_{25}$Sn$_4$	850, 8d	Ni$_3$Sn$_2$ HT Ni$_{12}$P$_5$ LT Ni$_{10}$P$_3$Sn T1	a=404.7(1) c=515.6(2) a=864.614(5) c=507.210(4) a=766.640(4) c=961.948(7)	not determined		
NPS 119	Ni$_{72}$P$_{24}$Sn$_4$	850, 8d	Ni$_5$P$_2$ LT T1 Ni$_{10}$P$_3$Sn T2 C$_6$Cr$_{23}$	a=661.74(2) c=1232.13(5) a=767.49(3) c=962.67(7) a=1111.64(2)	not determined		
NPS 120	Ni$_{73}$P$_{23}$Sn$_4$	850, 8d	Ni$_3$P T2 C$_6$Cr$_{23}$ T	a=895.64(3) c=438.84(2) a=1111.75(2) ---	not determined		
700 °C							
Ni$_2$PSn							
NPS 3	Ni$_{50}$P$_{25}$Sn$_{25}$	700, 52d	(Sn) (I) T5	a=583.10(9) b=318.20(8) a=1282.03(2) b=359.148(5) c=508.772(7)	2.6 48.6	0.0 25.2	97.4 26.2	101.6 101.2
NPS 3	Ni$_{63.33}$P$_{31.67}$Sn$_5$	700, 52d	Fe$_2$P Ni$_2$PSn	a=586.51(2) c=338.56(1) a=1282.57(5) b=359.65(1) c=509.21(2)	not determined		
NPS 4	Ni$_{60}$P$_{30}$Sn$_{10}$	700, 37d	Fe$_2$P Ni$_2$PSn	a=586.06(1) c=338.576(9) a=1282.07(1) b=359.463(4) c=508.951(6)	not determined		
NPS 5	Ni$_{56.67}$P$_{28.33}$Sn$_{15}$	700, 52d	Fe$_2$P Ni$_2$PSn	a=585.86(1) c=338.90(1) a=1282.08(2) b=359.325(4) c=508.868(7)	not determined		
NPS 6	Ni$_{53.33}$P$_{26.67}$Sn$_{20}$	700, 52d	Fe$_2$P Ni$_2$PSn	a=585.97(3) c=338.64(3) a=1281.76(2) b=359.347(4) c=508.861(6)	not determined		
NPS 7	Ni$_{46.67}$P$_{23.3}$Sn$_{30}$	700, 52d	βSn (I) Ni$_2$PSn	a=583.11(1) c=318.15(1) a=1282.19(1) b=359.252(3) c=508.854(5)	1.5 48.7	0.0 25.3	98.5 26.0	102.5 100.8
NPS 19	Ni$_{61.7}$P$_{8.3}$Sn$_{30}$	700, 44d	Ni$_3$Sn$_2$ HT Ni$_{12}$P$_5$ LT	a=406.93(1) c=515.74(1) a=864.42(7) c=506.77(5)	57.2 69.7	1.3 30.2	41.6 0.1	99.3 97.1

				Phase Analysis					
No.	Nominal Composition [at.%]	Heat Treatment [°C]	Phase	Structure Type	Lattice Param. [pm]	Ni	EDX / WDS [at.%] P	Sn	Σ wt.%
700 °C									
NPS 20	$Ni_{63}P_{17}Sn_{20}$	700, 44d	Ni_3Sn_2 HT	$InNi_2$	$a=404.751(5)\ c=514.824(9)$	56.9	2.3	40.9	101.5
			$Ni_{12}P_5$ LT	$Ni_{12}P_5$	$a=864.37(2)\ c=506.90(1)$	69.2	30.7	0.2	97.7
			T3	$Ni_{10}P_3Sn_5$	$a=646.0(1)\ b=841.83(9)\ c=1013.43(1)\ \alpha=75.30(2)°\ \beta=83.29(2)°\ \gamma=84.099(8)°$	56.1	16.1	27.8	100.4
NPS 21	$Ni_{65}P_{25}Sn_{10}$	700, 44d	$Ni_{12}P_5$ LT	$Ni_{12}P_5$	$a=864.64(3)\ c=506.98(2)$	69.5	30.4	0.1	98.4
			Ni_2P	Fe_2P	$a=586.43(8)\ c=339.13(7)$				
			T3	$Ni_{10}P_3Sn_5$	$a=646.02(3)\ b=842.31(4)\ c=1011.77(4)\ \alpha=75.343(4)\ \beta=83.349(4)\ \gamma=83.979(2)$	65.9	34.0	0.1	97.9
						55.8	16.3	27.9	101.3
NPS 22	$Ni_{72.92}P_{8.3}Sn_{18.75}$	700, 83d	Ni_3Sn LT	Mg_3Cd	$a=529.71(1)\ c=424.86(1)$	not determined			
			Ni_3Sn_2 HT	$InNi_2$	$a=413.325(8)\ c=520.69(2)$				
			T2	C_6Cr_{23}	$a=1111.67(2)$				
NPS 23	$Ni_{70.8}P_{16.7}Sn_{12.5}$	700, 83d	Ni_3Sn LT	Mg_3Cd	$a=529.79(3)\ c=424.94(5)$	76.6	0.0	23.4	99.0
			Ni_3Sn_2 HT	$InNi_2$	$a=413.487(6)\ c=520.735$	63.5	0.2	36.3	100.0
			T2	C_6Cr_{23}	$a=1111.73(2)$	73.2	20.0	6.9	99.4
NPS 24	$Ni_{68.75}P_{25}Sn_{6.25}$	700, 83d	Ni_3Sn_2 HT	$InNi_2$	$a=407.530(4)\ c=516.217(7)$	60.5	1.1	38.4	102.2
			$Ni_{12}P_5$ LT	$Ni_{12}P_5$	$a=864.859(6)\ c=507.207(4)$	71.0	28.8	0.2	100.0
NPS 25	$Ni_{71.25}P_5Sn_{23.75}$	700, 19d	Ni_3Sn LT	Mg_3Cd	$a=529.57(2)\ c=425.28(2)$	73.6	0.0	26.4	99.3
			Ni_3Sn_2 HT	$InNi_2$	$a=413.42(1)\ c=520.56(2)$	60.7	0.0	39.3	99.7
			T2	C_6Cr_{23}	$a=1111.72(3)$	71.2	21.4	7.4	97.7
NPS 26	$Ni_{67.5}P_{10}Sn_{22.5}$	700, 19d	Ni_3Sn LT	Mg_3Cd	$a=529.66(2)\ c=424.80(3)$	73.8	2.7	26.5	99.2
			Ni_3Sn_2 HT	$InNi_2$	$a=413.115(4)\ c=520.417(7)$	60.4	0.0	39.6	98.6
			T2	C_6Cr_{23}	$a=1111.40(1)$	70.8	21.6	7.6	97.7
NPS 27	$Ni_{63.75}P_{15}Sn_{21.25}$	700, 19d	Ni_3Sn_2 HT	$InNi_2$	$a=406.196(7)\ c=515.347(1)$	69.1	30.2	0.7	98.1
			$Ni_{12}P_5$ LT	$Ni_{12}P_5$	$a=864.55(2)\ c=507.03(2)$	57.5	2.1	40.5	99.4
NPS 28	$Ni_{60}P_{20}Sn_{20}$	700, 19d	$Ni_{12}P_5$ LT	$Ni_{12}P_5$	$a=864.49(3)\ c=506.97(2)$	69.7	30.1	0.2	98.3
			Ni_2P	Fe_2P	$a=585.92(9)\ c=339.15(8)$	not found in EPMA			
			T3	$Ni_{10}P_3Sn_5$	$a=647.00(3)\ b=842.88(3)\ c=1012.59(4)\ \alpha=75.304(4)°\ \beta=83.380(4)°\ \gamma=84.204(2)°$	55.3	15.4	29.3	99.8
NPS 29	$Ni_{56.25}P_{25}Sn_{18.75}$	700, 19d	Ni_2P	Fe_2P	$a=585.66(5)\ c=339.30(4)$	65.7	34.1	0.2	98.9
			T3	$Ni_{10}P_3Sn_5$	$a=641.31(1)\ b=833.41(1)\ c=1027.23(2)\ \alpha=73.364(1)°\ \beta=84.412(2)°\ \gamma=82.738(1)°$	54.1	16.9	29.0	100.8
			T5	Ni_2PSn	$a=1282.55(6)\ b=360.33(2)\ c=509.81(3)$	49.0	24.5	26.6	101.3

Sample	Composition	T(°C), time	Phase	Structure type	Lattice parameters				
NPS 30	Ni$_{52.5}$P$_{30}$Sn$_{17.5}$	700, 19d	Ni$_2$P	Fe$_2$P	a=585.7(1) c=338.9(1)	65.4	34.5	0.1	98.7
			Ni$_5$P$_4$	Ni$_5$P$_4$	a=679.88(2) c=1101.84(7)	54.5	44.6	0.9	98.3
			T5	Ni$_2$PSn	a=1280.34(1) b=358.598(4) c=508.357(6)	48.7	26.0	25.2	100.3
NPS 31	Ni$_{48.75}$P$_{35}$Sn$_{16.25}$	700, 14d	Sn (I)	βSn	a=583.270(7) c=318.253(5)	not determined			
			Ni$_5$P$_4$	Ni$_5$P$_4$	a=680.074(6) c=1102.60(1)				
			T5	Ni$_2$PSn	a=1279.7(3) b=358.59(7) c=508.9(1)				
NPS 32	Ni$_{45}$P$_{40}$Sn$_{15}$	700, 14d	Sn (I)	βSn	a=583.321(8) c=318.249(6)	not determined			
			Ni$_5$P$_4$	Ni$_5$P$_4$	a=679.865(6) c=1101.91(1)				
			NiP$_2$	NiP$_2$	a=636.92(2) b=561.68(1) c=607.31(1) β=126.240(2)°				
NPS 33	Ni$_{57}$P$_5$Sn$_{38}$	700, 14d	Ni$_3$Sn$_2$ HT	InNi$_2$	a=404.737(7) c=514.16(1)	not determined			
			T4	Ni$_{13}$P$_3$Sn$_8$	a=646.15(8) b=2136.7(3) c=1320.0(8) α=80.882(8)° β=59.35(1)° γ=68.11(1)°				
NPS 34	Ni$_{54}$P$_{10}$Sn$_{36}$	700, 19d	Ni$_3$Sn$_2$ HT	InNi$_2$	a=403.69(4) c=507.8(1)	55.5	0.5	44.0	99.6
			Ni$_3$Sn$_4$	Ni$_3$Sn$_4$	a=1245.01(6) b=408.201(6) c=520.83(2) β=103.611(4)	46.1	0.0	53.9	100.8
			Ni$_{10}$P$_3$Sn$_5$	Ni$_{10}$P$_3$Sn$_5$	a=646.3(1) b=841.3(1) c=1012.5(1) α=75.20(1)° β=83.52(2)° γ=84.05(1)°	54.4	15.7	29.9	100.2
			T4	Ni$_{13}$P$_3$Sn$_8$	a=647.93(4) b=2137.3(1) c=1318.11(7) α=80.761(4)° β=59.07(1)(5)° γ=67.945(4)°	54.0	13.5	32.5	99.5
NPS 35	Ni$_{51}$P$_{15}$Sn$_{34}$	700, 19d	Sn (I)	βSn	a=583.37(3) c=318.10(2)	1.5	0.0	98.5	101.3
			Ni$_3$Sn$_4$	Ni$_3$Sn$_4$	a=1223.6(3) b=406.40(9) c=513.7(1) β=101.62(2)°	41.6	0.0	58.4	100.3
			T3	Ni$_{10}$P$_3$Sn$_5$	a=???	54.0	16.7	29.3	99.8
NPS 36	Ni$_{48}$P$_{20}$Sn$_{32}$	700, 19d	Sn(I)	βSn	a=583.25(4) c=318.10(3)	2.8	0.0	97.2	101.8
			T3	Ni$_{10}$P$_3$Sn$_5$	a=650.53(3) b=833.23(4) c=994.81(4) α=77.638(4)° β=82.120(4)° γ=84.273(4)°	53.8	16.9	29.3	98.9
			T5	Ni$_2$PSn	a=1282.46(7) b=360.41(2) c=509.78(3)	48.8	24.4	26.8	99.0
NPS 39	Ni$_{39}$P$_{35}$Sn$_{26}$	700, 19d	Sn (I)	βSn	only qualitative XRD	0.6	0.0	99.4	98.7
			Ni$_5$P$_4$	Ni$_5$P$_4$		54.2	45.1	0.6	97.6
			NiP$_2$	NiP$_2$		30.3	65.8	3.9	98.3
			P$_3$Sn$_4$ (I)	Bi$_3$Se$_4$		0.8	41.0	58.2	102.1
NPS 40	Ni$_{36}$P$_{40}$Sn$_{24}$	700, 19d	Sn (I)	βSn	a=583.092(5) c=318.093(4)	0.4	0.0	99.6	101.5
			Ni$_5$P$_4$	Ni$_5$P$_4$	a=679.567(6) c=1101.21(1)	54.5	44.9	0.6	98.7
			NiP$_2$	NiP$_2$	a=644.9(2) b=571.5(2) c=629.4(2) β=126.57(2)°	32.8	67.2	0.0	99.1
			P$_3$Sn$_4$ (I)	Bi$_3$Se$_4$	traces only	not found in EPMA			

No.	Nominal Composition [at.%]	Heat Treatment [°C]	Phase Analysis			EDX / WDS [at.%]			Σ wt.%
			Phase	Structure Type	Lattice Param. [pm]	Ni	P	Sn	
700 °C									
NPS 41	$Ni_{42.75}P_5Sn_{52.25}$	700, 28d	(Sn) (I)	βSn	a= 583.43(2) c=318.18(1)	not found in EPMA			
			Ni_3Sn_4	Ni_3Sn_4	a=1237.82(3) b=407.20(1) c=521.46(1) β=104.091(3)°	45.7	0.0	54.7	101.5
			T3	$Ni_{10}P_3Sn_5$	a=647.72(5) b=838.09(5) c=1010.26(6) α=75.326(7)° β=83.456(7)° γ=84.090(5)°	55.9	15.5	28.6	101.7
NPS 42	$Ni_{40.5}P_{10}Sn_{49.5}$	700, 28d	(Sn) (I)	βSn	a= 583.36(2) c=318.12(1)	not determined			
			Ni_3Sn_4	Ni_3Sn_4	a=1226.5(3) b=408.11(8) c=519.5(1) b=104.68(2)°				
			T3	$Ni_{10}P_3Sn_5$	a=647.93(3) b=838.24(4) c=1010.08(5) α=75.320(4)° β=83.429(4)° γ=84.042(3)°				
NPS 43	$Ni_{38.25}P_{15}Sn_{46.75}$	700, 28d	(Sn) (I)	βSn	a=583.18(1) c=318.169(8)	not determined			
			Ni_3Sn_4 (I)	Ni_3Sn_4	a=1233.3(2) b=408.17(6) c=523.48(7) β=103.75(1)°.	not determined			
			T3	$Ni_{10}P_3Sn_5$	a=638.59(2) b=832.003(3) c=1031.19(3) α=74.192(3)° β=83.802(3)° γ=82.311(3)°	55.6	16.3	28.1	101.0
			T5	Ni_2PSn		50.3	23.7	25.9	100.9
NPS 48	$Ni_{27}P_{40}Sn_{33}$	700, 28d	(Sn) (I)	βSn	a=583.16(1) c=318.16(1)	not determined			
			Ni_5P_4	Ni_5P_4	a=679.75(1) c=1101.34(2)	55.5	43.9	0.6	100.4
			NiP_2	NiP_2	a=636.822(9) b=561.628(8) c=607.265(9) β=126.2356(7)°	33.3	66.7	0.1	99.8
NPS 49	$Ni_{73.9}P_{4.2}Sn_{21.9}$	700, 83d	Ni_3Sn LT	Mg_3Cd	a=529.98(1) c=425.09(1)	75.1	0.2	24.7	100.0
			Ni_3Sn_2 HT	$InNi_2$	a=413.27(4) c=528.4(1)	61.9	0.4	37.6	100.1
			T2	C_6Cr_{23}	a=1112.27(4)	71.0	22.1	6.8	Brmo*)
NPS 50	$Ni_{71.9}P_{12.5}Sn_{15.6}$	700, 83d	Ni_3Sn LT	Mg_3Cd	a=529.83(1) c=424.91(1)	75.0	0	25.0	
			Ni_3Sn_2 HT	$InNi_2$	a=413.45(1) c=520.74(2)	62.2	0.5	37.3	
			T2	C_6Cr_{23}	a=1111.75(2)	71.0	22.0	7.0	Brmo*)
NPS 51	$Ni_{69.8}P_{20.8}Sn_{9.4}$	700, 83d	Ni_3Sn_2 HT	$InNi_2$	a=407.645(5) c=516.280(9)	59.2	0.9	39.9	102.6
			$Ni_{12}P_5$ LT	$Ni_{12}P_5$	a=864.94(2) c=507.12(1)	70.5	29.4	0.1	100.2
			T1	$Ni_{10}P_3Sn$	a=767.719(6) c=961.89(1)	71.3	21.3	7.4	100.4
NPS 52	$Ni_{67.7}P_{29.2}Sn_{3.1}$	700, 83d	$Ni_{12}P_5$ LT	$Ni_{12}P_5$	a=864.71(1) c=507.08(8)	70.5	29.5	0.0	100.0
			Ni_2P	Fe_2P	a=586.48(2) c=339.22(2)	66.8	33.2	0.0	100.1
			T3	$Ni_{10}P_3Sn_5$	a=645.86(5) b=842.42(6) c=1011.66(7) α=75.319(8)° β=83.387(9)° γ=84.005(5)°	56.7	15.2	28.1	102.8
NPS 53	$Ni_{73.9}P_{4.2}Sn_{21.9}$	700, 83d	Ni_3Sn_2 HT	$InNi_2$	a=407.890(4) c=516.465(6)	59.9	1.7	38.4	101.5
			$Ni_{12}P_5$ LT	$Ni_{12}P_5$	a=887.4(4) c=496.3(3)	not found in EPMA			
			T1	$Ni_{10}P_3Sn$	a=768.09(4) c=961.14(9)	70.8	21.8	7.4	100.4

Sample	Composition	T (°C), time	Phases	Lattice parameters (pm, °)	Composition (at.%)			Total
NPS 54	$Ni_{71.9}P_{12.5}Sn_{15.6}$	700, 83d	Ni_3Sn_2 HT $Ni_{12}P_5$ LT T3	$a=405.031(3)$ $c=514.877(5)$ $a=864.70(1)$ $c=507.01(1)$ $a=647.66(7)$ $b=840.89(8)$ $c=1015.3(1)$ $\alpha=75.02(1)°$ $\beta=83.609(9)°$ $\gamma=85.00(1)°$	58.2 69.8 not found in EPMA	1.8 30.0	40.0 0.2	101.7 99.9
NPS 55	$Ni_{69.8}P_{20.8}Sn_{9.4}$	700, 83d	$InNi_2$ $Ni_{12}P_5$ $Ni_{10}P_3Sn_5$	$a=404.814(6)$ $c=514.89(1)$ $a=864.63(1)$ $c=506.89(1)$ $a=646.91(4)$ $b=842.47(4)$ $c=1012.34(5)$ $\alpha=75.270(6)°$ $\beta=83.350(6)°$ $\gamma=84.063(3)°$	58.9 70.0 57.0	2.4 29.8 15.4	38.6 0.2 27.6	102.1 99.7 101.7 Bmo[a]
NPS 56	$Ni_{67.7}P_{29.2}Sn_{3.1}$	700, 83d	$Ni_{12}P_5$ LT Ni_2P T3	$a=864.69(2)$ $c=507.10(1)$ $a=586.61(2)$ $c=339.09(1)$ $a=646.33(5)$ $b=842.41(5)$ $c=1011.91(6)$ $\alpha=75.284(7)°$ $\beta=83.349(8)°$ $\gamma=84.006(4)°$	70.1 66.3 57.2	29.9 33.6 15.4	0.0 0.0 27.4	100.6 100.4 102.5
NPS 63	$Ni_{90}P_5Sn_5$	700, 19d	(Ni) Ni_3Sn LT T2	$a=355.011(7)$ not found in XRD $a=1111.65(8)$	96.7 80.8 70.8	0.7 0.0 22.4	2.6 19.2 6.8	Bmo[a]
					systematic deviations in EDX measurements!			
NPS 64	$Ni_{80}P_5Sn_{15}$	700, 19d	(Ni) Ni_3Sn LT T2	$a=355.026(2)$ $a=529.30(1)$ $c=424.91(1)$ $a=1111.50(6)$	97.2 74.6 70.8	0.7 4.0 22.5	2.0 21.4 6.7	Bmo[a]
					systematic deviations in EDX measurements!			
NPS 65	$Ni_{90}P_{15}Sn_5$	700, 19d	(Ni) Ni_3Sn LT Ni_3P T2	$a=354.161(7)$ not found in XRD not found in XRD $a=1111.53(3)$	95.4 76.0 73.5 71.4	1.0 0.2 26.5 21.8	3.6 23.8 0.0 6.8	Bmo[a]
					systematic deviations in EDX measurements!			
NPS 79	$Ni_{71}P_{27}Sn_2$	700, 21d	Ni_5P_2 LT $Ni_{12}P_5$ LT T1	$a=661.74(4)$ $c=1228.8(2)$ $a=864.96(1)$ $c=507.023(9)$ $a=767.49(2)$ $c=962.40(4)$	70.6 not found in EPMA 71.6	28.1 21.1	1.3 7.2	101.3 100.6
NPS 79b	$Ni_{73}P_{25}Sn_2$	700, 21d	Ni_3P T1 $Ni_{10}P_3Sn$	$a=894.96(2)$ $c=438.63(1)$ $a=767.56(1)$ $c=961.30(3)$ ---	75.4 71.5 72.8	24.6 21.0 26.4	0.0 7.5 0.8	100.0 100.6 100.1
NPS 80	$Ni_{47}P_3Sn_{50}$	700, 30d	Ni_3Sn_4 $Ni_{10}P_3Sn_5$	$a=1236.30(3)$ $b=407.03(1)$ $c=521.43(1)$ $\beta=104.155(2)°$ $a=646.9(3)$ $b=838.9(3)$ $c=1010.1(3)$ $\alpha=75.31(4)°$ $\beta=83.50(5)°$ $\gamma=84.07(2)°$	45.0 55.9	0.0 15.6	55.0 28.5	101.2 101.3

No.	Nominal Composition [at.%]	Heat Treatment [°C]	Phase Analysis			EDX / WDS [at.%]			
			Phase	Structure Type	Lattice Param. [pm]	Ni	P	Sn	Σ wt.%
700 °C									
NPS 100	$Ni_{58}P_8Sn_{34}$	700, 17d	Ni_3Sn_2 HT	$InNi_2$	$a=403.944(8)$ $c=513.18(2)$	55.8	2.8	41.4	Bmo*)
			Ni_3Sn_4	Ni_3Sn_4	not found in XRD	46.8	1.3	51.9	
			T3	$Ni_{10}P_3Sn_5$	$a=646.41(8)$ $b=841.58(9)$ $c=1011.21(9)$ $α=75.26(1)°$ $β=83.40(1)°$ $γ=84.049(7)°$	54.0	17.2	28.9	
			T4	$Ni_{13}P_3Sn_8$	$a=647.537(7)$ $b=2151.2(1)$ $c=1322.54(1)$ $α=80.931(7)°$ $β=59.128(8)°$ $γ=67.757(8)°$	52.8	13.9	33.2	
NPS 101	$Ni_{57}P_8Sn_{35}$	700, 17d	Ni_3Sn_2 HT	$InNi_2$	$a=403.809(6)$ $c=512.81(1)$	55.2	1.9	42.9	Bmo*)
			Ni_3Sn_4	Ni_3Sn_4	$a=1244.43(7)$ $b=408.05(2)$ $c=521.03(3)$ $β=103.654(5)°$	45.3	0.0	54.7	
			T4	$Ni_{13}P_3Sn_8$	$a=648.14(6)$ $b=2135.6(1)$ $c=1319.0(1)$ $α=80.832(6)°$ $β=59.105(8)°$ $γ=68.004(8)°$	53.3	14.4	32.3	
			T3	$Ni_{10}P_3Sn_5$	not found in XRD	54.9	16.5	28.6	
NPS 102	$Ni_{56}P_8Sn_{36}$	700, 17d	Ni_3Sn_2 HT	$InNi_2$	$a=403.83(1)$ $c=512.79(3)$				
			Ni_3Sn_4	Ni_3Sn_4	$a=1243.42(6)$ $b=407.93(2)$ $c=521.11(3)$ $β=103.721(4)°$				
			T4	$Ni_{13}P_3Sn_8$	$a=647.85(3)$ $b=2131.78(7)$ $c=1319.69(5)$ $α=80.835(3)°$ $β=59.171(5)°$ $γ=68.080(3)°$				
T1	$Ni_{71.4}P_{21.4}Sn_{7.1}$	700, 21d	T1	$Ni_{10}P_3Sn$	$a=767.153(8)$ $c=962.42(2)$	not determined			
T2	$Ni_{72.4}P_{20.7}Sn_{6.9}$	700, 21d	T2	$Ni_{21}P_6Sn_2$	$a=1111.73(2)$	not determined			
550 °C									
Ni_2PSn									
NPS 1	$Ni_{50}P_{25}Sn_{25}$	550, 43d	T5	Ni_2PSn	$a=1282.31(2)$ $b=359.364(4)$ $c=508.922(7)$	not determined			
NPS 3	$Ni_{63.33}P_{31.67}Sn_5$	550, 53d	Ni_2P	Fe_2P	$a=585.88(3)$ $c=339.16(3)$	not determined			
			T5	Ni_2PSn	$a=1278.39(3)$ $b=360.056(9)$ $c=509.46(2)$				
NPS 4	$Ni_{60}P_{30}Sn_{10}$	550, 37d	Ni_2P	Fe_2P	$a=586.01(1)$ $c=338.74(1)$	not determined			
			T5	Ni_2PSn	$a=1279.01(3)$ $b=359.912(9)$ $c=509.27(1)$				
NPS 5	$Ni_{56.67}P_{28.33}Sn_{15}$	550, 37d	Ni_2P	Fe_2P	$a=585.87(1)$ $c=338.897(9)$	65.8	34.2	0.0	98.3
			T5	Ni_2PSn	$a=1282.08(1)$ $b=359.327(4)$ $c=508.873(6)$	49.3	24.9	25.8	102.7
NPS 6	$Ni_{53.33}P_{26.67}Sn_{20}$	550, 37d	Ni_2P	Fe_2P	$a=585.47(2)$ $c=338.93(2)$	66.1	33.9	0.0	99.1
			T5	Ni_2PSn	$a=1277.94(4)$ $b=359.86(1)$ $c=509.19(2)$	49.0	25.2	25.8	101.5
NPS 8	$Ni_{44.33}P_{21.67}Sn_{35}$	550, 48d	Sn (l)	βSn	$a=583.09(1)$ $c=318.10(1)$ solidified from liquid	0.7	0.0	99.3	102.2
			P_3Sn_4 (l)	Bi_3Se_4	not found in XRD	0.5	41.0	58.5	100.9
			T5	Ni_2PSn	$a=1282.27(2)$ $b=359.270(5)$ $c=508.792(8)$	48.5	25.4	26.1	98.6

Sample	Composition	T, t	Phase	Lattice parameters	EPMA Ni	EPMA P	EPMA Sn	Total
NPS 9	Ni$_{40}$P$_{20}$Sn$_{40}$	550, 48d	βSn	a=583.103(6) c=318.107(5) solidified from liquid	0.0	0.0	100.0	100.9
			P$_3$Sn$_4$ (l)	not found in XRD	0.4	40.7	58.9	101.4
			T5	a=1282.30(1) b=359.268(4) c=508.825(6)	48.6	25.0	26.4	100.8
NPS 10	Ni$_{36.67}$P$_{18.33}$Sn$_{45}$	550, 48d	βSn	a=583.04(1) c=318.10(1) solidified from liquid	not determined			
			T5	a=1282.02(2) b=359.26(1) c=508.76(1)				
NPS 12	Ni$_{26.67}$P$_{13.33}$Sn$_{60}$	550, 48d	βSn	no XRD made	0.8	0.0	99.2	102.9
			Ni$_2$PSn		48.1	25.6	26.3	100.5
NPS 13*)	Ni$_{23.33}$P$_{11.67}$Sn$_{65}$	550, 59d	βSn	a=583.223(3) c=318.254(2) solidified from liquid	not determined			
			T5	a=1282.20(2) b=359.367(6) c=508.91(1)				
			Ni$_2$PSn +unind. lines					
NPS 19	Ni$_{61.67}$P$_{8.33}$Sn$_{30}$	550, 59d	InNi$_2$	a=406.37(1) c=516.08(1)	58.0	3.6	38.4	99.5
			Ni$_{12}$P$_5$	a=864.42(6) c=506.844)	Too small to be measured			
NPS 20	Ni$_{63.33}$P$_{16.67}$Sn$_{20}$	550, 59d	InNi$_2$	a=405.40(1) c=514.61(2)	56.6	4.6	38.8	100.5
			Ni$_{12}$P$_5$ LT	a=864.72(3) c=507.08(2)	69.4	30.3	0.3	98.8
			T4	a=650.1(1) b=2193.8(3) c=1339.8(2) α=80.67(1)° β=58.79(2)° γ=67.72(2)°	55.6	13.8	30.6	100.5
NPS 21	Ni$_{65}$P$_{25}$Sn$_{10}$	550, 59d	Ni$_{12}$P$_5$ LT	a=864.4(2) c=507.18(7)	69.4	30.4	0.2	98.5
			Ni$_2$P	a=587.3(3) c=338.4(8)				
			T3	a=648.02(6) b=846.63(8) c=1023.77(6) α=74.982(7) β=83.581(7) γ=84.794(5)	55.8	14.5	29.7	101.7
NPS 22	Ni$_{72.92}$P$_{8.33}$Sn$_{18.75}$	550, 169d	Ni$_3$Sn LT	a=529.72(1) c=424.90(1)	not determined			
			Ni$_3$Sn$_2$ HT	a=412.78(1) c=520.30(3)				
			T2	a=1111.81(2)				
NPS 23	Ni$_{70.83}$P$_{16.67}$Sn$_{12.5}$	550, 169d	Ni$_3$Sn LT	a=529.52(4) c=425.25(7)	not found in EPMA			
			InNi$_2$	a=412.845(7) c=520.29(1)	63.7	0.4	35.9	99.4
			C$_6$Cr$_{23}$	a=1111.50(1)	73.2	20.0	6.8	100.1
NPS 24	Ni$_{68.75}$P$_{25}$Sn$_{6.25}$	550, 169d	InNi$_2$	a=406.793(6) c=516.08(1)	61.8	2.6	35.5	99.5
			Ni$_{12}$P$_5$ LT	a=864.708(9) c=507.219(6)	71.0	28.8	0.2	100.2
NPS 25	Ni$_{71.25}$P$_{5}$Sn$_{23.75}$	550, 19d	Ni$_3$Sn LT	a=529.60(1) c=424.75(1)	73.1	1.8	25.1	100.9
			InNi$_2$	a=412.83(2) c=520.23(2)	61.3	2.0	36.7	101.3
			C$_6$Cr$_{23}$	a=1111.57(4)	71.4	21.3	7.4	99.8
NPS 26	Ni$_{67.5}$P$_{10}$Sn$_{22.5}$	550, 19d	Ni$_3$Sn LT	a=529.80(4) c=424.92(6)	71.3	21.1	7.6	100.0
			Ni$_3$Sn$_2$ HT	a=412.75(1) c=520.20(2)	60.8	1.3	37.9	101.3
			C$_6$Cr$_{23}$	a=1111.58(2)	71.4	21.3	7.4	99.8

No.	Nominal Composition [at.%]	Heat Treatment [°C]	Phase	Structure Type	Lattice Param. [pm]	EDX / WDS [at.%] Ni	P	Sn	Σ wt.%
550 °C									
NPS 27	$Ni_{63.75}P_{15}Sn_{21.25}$	550, 19d	Ni_3Sn_2 HT	$InNi_2$	a=405.15(2) c=515.46(3)	57.4	3.3	39.2	101.3
			$Ni_{12}P_5$ LT	$Ni_{12}P_5$	a=864.62(3) c=507.17(2)	69.3	30.5	0.2	98.6
			T4	$Ni_{13}P_3Sn_8$	a=653.2(2) b=2204.6(3) c=1344.2(3) α=82.46(2)° β=58.47(2)° γ=67.86(2)°	not found in EPMA			
NPS 28	$Ni_{60}P_{20}Sn_{20}$	550, 19d	Ni_2P	Fe_2P	a=585.32(5) c=338.76(4)	65.1	34.9	0.0	97.0
			$Ni_{12}P_5$ LT	$Ni_{12}P_5$	a=863.97(9) c=506.50(6)				
			T3	$Ni_{10}P_3Sn_5$	a=647 b=842 c=1010.8 α=75.3° β=83.35° γ=84.05°	55.0	14.1	30.9	100.2
NPS 29	$Ni_{56.25}P_{25}Sn_{18.75}$	550, 19d	Ni_2P	Fe_2P	a=585.70(5) c=388.84(4)	65.4	64.5	0.1	98.2
			T3	$Ni_{10}P_3Sn_5$	a=640.58(3) b=833.41(3) c=1026.98(4) α=73.416(3) β=84.411(4) γ=82.738(3)	54.2	16.6	29.2	100.2
			T5	Ni_2PSn	a=1277.5(1) b=361.25(3) c=510.13(5)	48.8	24.7	26.5	100.3
NPS 30	$Ni_{52.5}P_{30}Sn_{17.5}$	550, 19d	Ni_5P_4	Ni_5P_4	a=680.04(5) c=1101.6(1)	54.8	44.6	0.6	98.1
			Ni_2P	NiP_2	a=585.5(2) c=339.64(1)	65.3	34.7	0.0	97.8
			T5	Ni_2PSn	a=1280.76(4) b=329.37(1) c=508.97(2)	49.0	25.2	25.8	100.7
NPS 31	$Ni_{48.75}P_{35}Sn_{16.25}$	550, 30d	(Sn) (l)	βSn	a=583.32(1) c=318.239(9) solidified from liquid	not determined			
			Ni_5P_4	Ni_5P_4	a=679.70(3) c=1102.70(7)				
			T5	Ni_2PSn	a=1281.48(3) b=359.005(8) c=508.78(1)				
NPS 32	$Ni_{45}P_{40}Sn_{15}$	550, 30d	(Sn) (l)	βSn	a=583.31(2) c=318.37(2) solidified from liquid	not determined			
			Ni_5P_4	Ni_5P_4	a=679.89(2) c=1101.56(5)				
			NiP_2	NiP_2	a=636.95(2) b=561.65(2) c=607.32(2) β=126.241(2)°				
			T5	Ni_2PSn	a=1281.63(2) b=38.938(6) c=508.730(9) only small amount				
NPS 33	$Ni_{57.5}Sn_{38}$	550, 59d	Ni_3Sn_2 HT	$InNi_2$	a=405.004(7) c=514.34(1)	not determined			
			T4	$Ni_{13}P_3Sn_8$	a=649.89(8) b=2195.2(2) c=1349.4(1) α=81.565(7)° β=58.97(1)° γ=67.67(1)°				
NPS 34	$Ni_{54}P_{10}Sn_{36}$	550, 19d	Ni_3Sn_2 HT	$InNi_2$	a=409.437(5) c=513.039(10)	not found in EPMA			
			Ni_3Sn_4	Ni_3Sn_4	a=1223.9(2) b=407.09(6) c=522.13(8) β=104.04(1)°	45.2	0.0	54.8	101.1
			T4	$Ni_{13}P_3Sn_8$	a=646.10(7) b=2149.8(1) c=1323.3(1) α=80.849(5)° β=59.17(1)° γ=67.96(1)°	54.2	11.8	34.0	99.9
NPS 35	$Ni_{51}P_{15}Sn_{34}$	550, 19d	(Sn) (l)	βSn	a=583.08(4) c=318.07(2) solidified from liquid	2.8	0.0	97.2	101.9
			Ni_3Sn_4	Ni_3Sn_4	a=1261.4(2) b=407.82(6) c=503.71(8) β=103.52(1)°	41.7	0.0	58.3	100.0
			T3	$Ni_{10}P_3Sn_5$	a=646.60(7) b=840.31(8) c=1008.6(7) α=75.324(6)° β=83.409(7)° γ=83.926(4)°	54.6	16.3	29.1	100.5
			T5	Ni_2PSn	a=1274.8(4) b=360.4(1) c=513.6(2)	48.9	24.4	26.7	99.7

Sample	Cond.	Col3	Col4	Lattice parameters	V1	V2	V3	V4
NPS 36	550, 19d	Sn (l)	βSn	a=583.18(2) c=318.08(2) solidified from liquid	2.6	0.0	97.5	102.6
		T3	Ni₁₀P₃Sn₅	a=645.47(4) b=840.96(5) c=1009.00(5) α=75.327(6) β=83.356(7) γ=83.875(4)	53.9	16.8	29.3	100.7
		T5	Ni₂PSn	a=1281.71(6) b=360.490(1) c=509.782(2)	48.5	24.7	26.8	101.3
NPS 37	550, 19d	Sn (l)	βSn	a=582.992(9) c=318.052(7) solidified from liquid	not determined			
		NiP₂	NiP₂	a=636.47(4) b=516.63(4) c=607.11(4) β=126.237(4)°				
		T5	Ni₂PSn	a=1281.46(2) b=359.024(4) c=508.587(7)				
NPS 38	550, 19d	Sn (l)	βSn	a=582.951(7) c=318.030(5) solidified from liquid	1.8	0.0	98.2	102.2
		NiP₂	NiP₂	a=636.55(6) b=561.40(5) c=607.14(5) β=126.245(5)°	32.4	67.5	0.1	99.1
		T5	Ni₂PSn	a=1281.13(2) b=358.894(4) c=508.502(7)	48.4	25.6	26.0	101.2
NPS 39	550, 19d	Sn (l)	βSn	a=583.099(8) c=318.111(6) solidified from liquid	2.32	0.0	97.7	104.6
		NiP₄	NiP₄	a=678.3(2) c=1109.5(5)	54.7	45.0	0.3	99.8
		NiP₂	NiP₂	a=636.69(1) b=561.47(1) c=607.10(1) β=126.232(1)°	32.1	67.9	0.0	98.8
		T5	Ni₂PSn	a=1281.15(2) b=358.919(5) c=508.587(8)	48.3	25.7	26.0	98.8
NPS 40	550, 19d	Sn (l)	βSn	a=583.100(9) c=318.117(6) solidified from liquid	1.4	0.0	98.6	101.7
		NiP₂	NiP₂	a=636.61(1) b=561.45(1) c=607.07(1) β=126.23°	32.5	67.4	0.1	98.6
		T5	Ni₂PSn	a=1281.64(3) b=359.032(8) c=508.65(1) + a few unindexed lines	48.6	25.5	25.9	100.6
					47.7	35.0	17.3	99.6
NPS 41	550, 28d	(Sn) (l)	βSn	a=583.32(1) c=318.16(1) solidified from liquid	not determined			
		Ni₃Sn₄	Ni₃Sn₄	a=1226.89(2) b=406.35(1) c=521.92(1) β=104.800(1)°				
		T3	Ni₁₀P₃Sn₅	a=645.3(1) b=841.1(1) c=1010.6(1) α=75.37(2)° β=83.33(2)° γ=83.96(1)°				
NPS 42	550, 28d	(Sn) (l)	βSn	a=583.19(1) c=318.21(1) solidified from liquid	not determined			
		Ni₃Sn₄	Ni₃Sn₄	a=1225.8(1) b=406.37(3) c=521.70(4) β=104.902(6)°				
		T3	Ni₁₀P₃Sn₅	a=645.78(4) b=841.50(4) c=1009.51(5) α=75.347(5)° β=83.832(5)° γ=83.905(3)°				
NPS 43	550, 28d	(Sn) (l)	βSn	a=583.26(1) c=318.066(8) solidified from liquid	not determined			
		T3	Ni₁₀P₃Sn₅	a=641.48(9) b=830.5(1) c=1020.1(1) α=74.78(1)° β=84.51(1)° γ=82.387(9)°				
		T5	Ni₂PSn	a=1282.39(4) b=360.21(1) c=509.69(2)				
NPS 48	550, 28d	(Sn) (l)	βSn	a=583.18(1) c=318.222(8) solidified from liquid	not determined			
		NiP₂	NiP₂	a=636.86(1) b=561.622(9) c=607.34(1) β=126.24(9)				
		T5	Ni₂PSn	a=1283.0(1) b=358.88(2) c=508.81(4) + some unindexed lines				
NPS 49	550, 169d	Ni₃Sn LT	Mg₃Cd	a=529.63(1) c=424.78(1)	not determined			
		Ni₃Sn₂ HT	InNi₂	a=406.82(7) c=533.1(2)				
		T2	C₆Cr₂₃	a=1111.94(5)				

No.	Nominal Composition [at.%]	Heat Treatment [°C]	Phase	Structure Type	Lattice Param. [pm]	Ni	EDX / WDS [at.%] P	Sn	Σ wt.%
550 °C									
NPS 50	$Ni_{71.88}P_{12.5}Sn_{15.63}$	550, 169d	Ni_3Sn LT Ni_3Sn_2 HT T2	Mg_3Cd $InNi_2$ C_6Cr_{23}	a=529.69(1) c=424.87(1) a=412.95(1) c=520.36(2) a=1111.85(2)	76.9 63.1 73.0	0.1 0.0 19.5	23.0 36.9 7.5	96.8 99.6 100.3
NPS 51	$Ni_{69.79}P_{20.83}Sn_{9.38}$	550, 169d	Ni_3Sn_2 HT $Ni_{12}P_5$ LT T1	$InNi_2$ $Ni_{12}P_5$ $Ni_{10}P_3Sn$	a=407.354(6) c=515.84(1) a=864.66(2) c=507.12(2) a=766.935(8) c=962.40(1)	61.2 71.3 72.1	2.7 27.1 21.4	36.1 1.5 6.5	100.1 99.6 100.1
NPS 52	$Ni_{67.71}P_{29.17}Sn_{3.13}$	550, 169d	$Ni_{12}P_5$ LT Ni_3P T3	$Ni_{12}P_5$ Fe_2P $Ni_{10}P_3Sn_5$	a=864.71(1) c=507.149(7) a=586.95(2) c=338.84(2) a=646.72(5) b=842.06(6) c=1010.94(6) α=75.349(6)° β=83.426(8)° γ=84.052(5)°	71.1 67.5 58.3	28.9 32.5 15.2	0.1 0.1 26.5	100.0 99.9 101.5
NPS 53	$Ni_{60.83}P_{4.17}Sn_{35}$	550, 169d	Ni_3Sn_2 HT $Ni_{12}P_5$ LT T1	$InNi_2$ $Ni_{12}P_5$ $Ni_{10}P_3Sn$	a=407.946(4) c=516.903(7) a=885.3(2) c=527.6(2) (traces only) a=768.33(8) c=961.7(1)	61.3 Not found in EPMA 71.4	2.4 19.8	36.3 8.8	98.5 100.5
NPS 54	$Ni_{62.5}P_{12.5}Sn_{25}$	550, 169d	Ni_3Sn_2 HT $Ni_{12}P_5$ LT $Ni_{13}P_3Sn_8$	$InNi_2$ $Ni_{12}P_5$ $Ni_{13}P_3Sn_8$	a=405.856(5) c=514.643(8) a=864.58(2) c=507.13(1) traces only	59.0 71.2	2.3 28.6	38.8 0.2	99.6 98.6
NPS 55	$Ni_{64.17}P_{20.83}Sn_{15}$	550, 169d	Ni_3Sn_2 HT $Ni_{12}P_5$ LT T4	$InNi_2$ $Ni_{12}P_5$ $Ni_{13}P_3Sn_8$	a=405.82(1) c=514.71(2) a=864.76(2) c=507.17(1) a=648.68(4) b=2166.73(9) c=1336.29(6) α=80.876(4)° β=58.974(5)° γ=67.881(6)°	59.1 70.8 58.2	0.7 28.9 11.0	40.2 0.4 30.8	101.4 99.7 100.3
NPS 56	$Ni_{65.83}P_{29.17}Sn_5$	550, 169d	$Ni_{12}P_5$ LT Ni_3P T3	$Ni_{12}P_5$ Fe_2P $Ni_{10}P_3Sn_5$	a=864.68(1) c=507.04(1) a=586.81(2) c=338.79(1) a=647.05(6) b=842.01(7) c=1010.81(7) α=75.277(8)° β=83.346(9)° γ=84.054(5)°	71.2 67.4 57.5	28.8 32.5 15.4	0.1 0.1 27.2	99.6 100.2 101.0
NPS 62	$Ni_{74.5}P_{23}Sn_{2.5}$	550, 59d	(Ni) Ni_3P T2	Cu Ni_3P C_6Cr_{23}	a=353.83(2) a=895.597(8) c=438.879(4) a=1111.78(1)	97.0 75.1 72.4	1.4 24.8 20.3	1.7 0.0 7.3	99.6 100.0 100.7
NPS 64	$Ni_{80}P_5Sn_{15}$	550, 59d	(Ni) Ni_3Sn LT T2	Cu Mg_3Cd C_6Cr_{23}	a=353.91(2) a=529.37(2) c=424.96(2) a=1111.38(8)	98.1 76.5 75.0	0.2 1.7 18.5	1.7 21.7 6.5	100.4 99.5 100.3
NPS 65	$Ni_{80}P_{15}Sn_5$	550, 59d	(Ni) Ni_3P T2	Cu Ni_3P C_6Cr_{23}	a=354.144(6) a=895.27(8) c=439.12(6) a=1111.57(2)	97.0 75.5 73.3	0.8 24.4 20.1	2.2 0.1 6.6	100.1 100.0 100.4

Sample	Conditions	Phase	Lattice parameters				
NPS 70	550, 34d	(Sn)	a=583.12(2) c=318.07(1)	0.0	0.0	100.0	99.9
		NiP$_2$	traces only	33.9	66.1	0.0	99.7
		NiP$_3$	a=782.25(2)	25.6	74.2	0.2	99.6
		P$_3$Sn$_4$	a=396.82(3) c=3534.4(5)	0.0	41.0	59.0	100.3
NPS 72	550, 34d	(Sn)	a=583.19(1) c=318.153(8)	0.0	0.0	100.0	99.2
		NiP$_3$	a=782.21(1)	25.0	74.9	0.1	99.4
		Bi$_3$Se$_4$	a=396.89(2) c=3534.4(3)	0.1	41.1	58.9	100.5
NPS 75	550, 34d	(Sn) (l)	No XRD made	0.1	0.0	100.0	100.1
		Ni$_3$P$_4$		56.0	43.5	0.5	99.3
		NiP$_2$		33.7	66.3	0.0	99.2
		P$_3$Sn$_4$		0.0	40.5	59.5	100.4
		T5		50.0	24.1	25.9	100.7
NPS 77	550, 34d	(Sn) (l)	a=583.11(2) c=318.12(1)	0.0	0.0	100.0	100.6
		NiP$_2$	a=636.85(1) b=561.630(8) c=607.20(1) β=126.1255(8)°	34.0	66.0	0.0	99.3
		P$_3$Sn$_4$	a=396.82(5) c=3533.7(9)	0.0	41.1	58.9	100.3
NPS 78	550, 34d	(Sn) (l)	a=583.14(1) c=318.10(1)	0.0	0.0	100.0	100.5
		NiP$_2$	a=637.05(1) b=561.59(1) c=607.35(1) β=126.252(1)°	34.1	65.9	0.0	99.3
		NiP$_3$	a=782.089(8)	25.7	74.3	0.1	98.8
		Bi$_3$Se$_4$	a=396.74(2) c=3534.9(5)	0.6	40.9	58.5	100.9
NPS 79	550, 24d	Ni$_5$P$_2$ LT	a=661.65(3) c=1224.5(1)	70.7	28.6	0.7	100.5
		Ni$_{12}$P$_5$ LT	a=864.634(9) c=507.166(6)	70.2	29.7	0.1	100.3
		T1	a=766.83(1) c=962.24(2)	71.2	21.6	7.2	100.9
NPS 79b	550, 24d	Ni$_3$P	a=894.84(8) c=438.91(5)	74.9	25.1	0.0	100.0
		Ni$_{10}$P$_3$Sn	a=768.08(5) c=960.74(9)	71.4	21.1	7.5	100.8
		T	---	72.8	26.4	0.8	100.2
NPS 93[e)]	550, 71d	(Sn) (l)	a=583.32(3) c=318.20(2)	not determined			
		NiP$_3$	a=782.29(1)				
		Bi$_3$Se$_4$	a=396.874(8) c=3535.4(1)				
NPS 94[e)]	550, 71d	(Sn) (l)	traces only	not determined			
		NiP$_3$	a=782.46(3)				
		Bi$_3$Se$_4$	a=396.93(1) c=3535.3(1)				
NPS 95[e)]	550, 71d	(Sn) (l)	a=583.23(5) c=318.20(4)	not determined			
		NiP$_2$	a=637.28(7) b=561.57(6) c=607.65(6) β=126.26(7)°				
		NiP$_3$	a=782.26(1)				
		Bi$_3$Se$_4$	a=396.812(6) c=3534.24(7)				

No.	Nominal Composition [at.%]	Heat Treatment [°C]	Phase	Structure Type	Lattice Param. [pm]	EDX / WDS [at.%]			
						Ni	P	Sn	Σ wt.%
550 °C									
NPS 96[c]	$Ni_{21.3}P_{57.4}Sn_{21.3}$	550, 71d	(Sn) (l)	βSn	a=583.35(1) c=318.317(7)	not determined			
			NiP_2	NiP_2	a=637.34(6) b=561.50(5) c=608.22(6) β=126.31(6)°				
			NiP_3	NiP_3	a=782.258(7)				
			P_3Sn_4	Bi_3Se_4	a=396.90(4) c=3535.2(8)				
NPS 97[c]	$Ni_{25.8}P_{58.7}Sn_{15.5}$	550, 71d	(Sn) (l)	βSn	a=583.167(5) c=318.158(5)	not determined			
			NiP_2	NiP_2	a=636.858(8) b=561.638(6) c=607.286(7) β=126.236(7)°				
			NiP_3	NiP_3	a=782.086(4)				
			P_3Sn_4	Bi_3Se_4	a=396.80(5) c=3534.6(8)				
NPS 98[c]	$Ni_{31}P_{58.7}Sn_{10.3}$	550, 71d	(Sn) (l)	βSn	a=583.199(6) c=318.160(5)	not determined			
			NiP_2	NiP_2	a=636.836(7) b=561.638(6) c=607.277(6) β=126.2360(6)°				
NPS 99[c]	$Ni_{36.2}P_{58.7}Sn_{5.1}$	550, 71d	(Sn) (l)	βSn	a=583.22(1) c=318.17(1)	not determined			
			NiP_2	NiP_2	a=636.86(1) b=561.659(8) c=607.283(9) β=126.2354(7)°				
			T5	Ni_2PSn	a=1281.56(8) b=358.86(2) c=508.66(3)				
200 °C									
NPS 14[*]	$Ni_{20}P_{10}Sn_{70}$	200, 125d	(Sn)	βSn	a=583.29(1) c=318.30(1)	not determined			
			T5	Ni_2PSn	a=1280.5(5) b=360.0(1) c=509.4(2)				
NPS 15[*]	$Ni_{13.33}P_{6.67}Sn_{80}$	200, 125d	(Sn)	βSn	a=583.367(5) c=318.295(3)	not determined			
			Ni_2P	Fe_2P	a=586.37(1) c=338.86(1)				
			T5	Ni_2PSn	a=1248.3(4) b=365.1(1) c=526.8(2)				
NPS 16[*]	$Ni_{10}P_5Sn_{85}$	200, 125d	(Sn)	βSn	a=583.168(6) c=318.183(4)	not determined			
			T5	Ni_2PSn	a=1276.1(3) b=360.32(8) c=509.4(1)				
NPS 17[*]	$Ni_{6.67}P_{3.33}Sn_{90}$	200, 125d	(Sn)	βSn	a=583.312(7) c=318.269(5)	not determined			
			T5	Ni_2PSn	a=1280.5(5) b=359.9(1) c=508.9(2)				
NPS 18[*]	$Ni_{3.33}P_{1.67}Sn_{95}$	200, 125d	(Sn)	βSn	a=583.22(1) c=318.230(6)	not determined			
			T5	Ni_2PSn	traces only				
NPS 66[*]	$Ni_5P_5Sn_{90}$	200, 102d	(Sn)	βSn	a=583.126(2) c=318.154(2)	1.0	0.0	99.6	101.3
			Ni_2P	Fe_2P	not found in XRD	67.0	32.9	0.0	100.0
			Ni_5P_4	Ni_5P_4		55.5	44.0	0.5	99.8
			NiP_2	NiP_2	a=638(3) b=560(2) c=590(3) β=125.5(3)° very low amount	33.8	66.3	0.0	99.7
			T5	Ni_2PSn	a=1279.7(8) b=359.9(2) c=507.9(3)	49.8	25.0	25.1	100.9

Sample	Anneal	Phase	Lattice parameters (pm)				
NPS 67[*)]	200, 102d	(Sn) NiP$_2$ P$_3$Sn$_4$	a=583.264(4) c=318.216(3) a=637.7(4) b=562.0(3) c=606.2(3) β=126.13(3)° a=396.78(5) c=3534(1)				not determined
NPS 68[*)]	200, 102d	(Sn) NiP$_2$ P$_3$Sn$_4$	a=583.170(5) c=318.193(4) a=632.3(7) b=561.6(6) c=603.9(6) β=125.79(7)° a=396.85(3) c=3533.3(7)				not determined
NPS 69[*)]	200, 102d	(Sn) NiP$_2$ P$_3$Sn$_4$	a=583.175(4) c=318.190(3) a=636.9(2) b=560.3(1) c=607.3(1) β=126.26(1)° a=396.85(3) c=3533.3(7)	0.0 33.7 0.0	0.0 66.3 41.1	100.0 0.0 58.9	100.1 100.3 100.4
NPS 70[*)]	200, 102d	(Sn) NiP$_2$ P$_3$Sn$_4$	a=583.186(4) c=318.197(3) a=638.7(4) b=560.6(3) c=607.7(3) β=126.30(3)° a=396.902(3) c=3534.55(5)				not determined
NPS 71[*)]	200, 102d	(Sn) NiP$_2$ P$_3$Sn$_4$	a=583.172(9) c=318.200(7) a=639.9(2) b=560.8(1) c=606.1(1) β=125.90(1)° a=396.899(4) c=3534.36(7)				not determined
NPS 72[*)]	200, 102d	(Sn) NiP$_2$ P$_3$Sn$_4$	a=583.25(1) c=318.22(1) a=637.0(1) b=560.19(7) c=607.69(9) β=126.232(9)° a=396. 91(4) c=3534.4(3)				not determined
NPS 73[*)]	200, 102d	(Sn) NiP$_2$ P$_3$Sn$_4$	a=583.25(3) c=318.17(2) a=634.5(2) b=558.3(2) c=609.0(2) β=125.83(2)° a=397.01(1) c=3534.7(2)				not determined
NPS 74[*)]	200, 102d	(Sn) Ni$_3$Sn$_4$ Ni$_2$P Ni$_2$PSn	a=583.157(6) c=318.200(4) a=1219.3(1) b=406.02(4) c=522.08(5) β=105.218(8)° not found in XRD a=1277.6(6) b=361.5(2) c=509.8(2)	0.0 42.3 62.6 47.4	0.0 0.2 37.4 27.1	100.0 57.5 0.0 25.5	Brno[*)]
NPS 75[*)]	200, 102d	(Sn) Fe$_2$P NiP$_2$ Ni$_2$PSn	a=583.367(4) c=318.270(3) a=586.395(4) c=338.92(4) a=633(1) b=562.8(9) c=603.6(9) β=125.8(1)° very low amount a=1279.9(3) b=359.22(7) c=508.8(1)	0.3 62.6 30.1 47.1	0.5 37.3 69.7 29.1	99.2 0.1 0.2 23.8	Brno[*)]
NPS 76[*)]	200, 102d	(Sn) Ni$_5$P$_4$ NiP$_2$ P$_3$Sn$_4$ Ni$_2$PSn	a=583.311(4) c=318.279(3) a=694.4(2) c=1109.1(5) a=636.87(6) b=561.59(5) c=607.34(6) β=126.235(6)° not found in XRD a=1280.1(7) b=359.4(2) c=509.2(3) unidentified	0.0 30.0 0.0 46.2 46.5	0.0 not found in EPMA 70.0 46.5 29.2 40.6	100.0 0.0 53.5 24.6 13.0	Brno[*)]

No.	Nominal Composition [at.%]	Heat Treatment [°C]	Phase	Structure Type	Lattice Param. [pm]	Ni	EDX / WDS [at.%] P	Sn	Σ wt.%
200 °C									
NPS 77[*)]	$Ni_{15}P_{35}Sn_{50}$	200, 102d	(Sn)	βSn	a=583.204(5) c=318.245(3)	0.0	0.0	100.0	Brno[*)]
			NiP_2	NiP_2	a=636.81(3) b=561.56(3) c=607.33(3) b=126.233(3)°	30.3	69.6	0.1	
			P_3Sn_4	Bi_3Se_4	a=396.868(7) c=3534.5(1)	0.0	45.9	54.1	
NPS 78[*)]	$Ni_{15}P_{40}Sn_{45}$	200, 102d	(Sn)	βSn	a=583.224(5) c=318.230(3)	not found in EPMA			Brno[*)]
			NiP_2	NiP_2	a=636.23(4) b=561.70(3) c=606.62(4) b=126.170(4)°	30.7	69.3	0.0	
			P_3Sn_4	Bi_3Se_4	a=396.905(7) c=3534.6(1)	0.2	46.3	53.5	
NPS 82[*)]	$Ni_{30}P_5Sn_{65}$	200, 64d	(Sn)	βSn	a=582.78(5) c=317.99(4)	not determined			
			Ni_3Sn_4	Ni_3Sn_4	a=1219.6(2) b=405.819(8) c=521.99(1) β=105.237(2)°				
			T3	$Ni_{10}P_3Sn_5$	a=647.9(2) b=840.2(2) c=1009.0(2) α=75.35(2)° β=83.34(3)° γ=83.88(2)°				
NPS 83[*)]	$Ni_{30}P_{10}Sn_{60}$	200, 64d	(Sn)	βSn	a=583.178(3) c=318.259(3)	not determined			
			Ni_3Sn_4	Ni_3Sn_4	a=1219.6(2) b=405.819(8) c=521.99(1) β=105.237(2)°				
			T5	Ni_2PSn	a=1281.45(6) b=360.50(2) c=510.03(3)				
NPS 84[*)]	$Ni_{30}P_{15}Sn_{55}$	200, 64d	(Sn)	βSn	a=583.21(2) c=318.249(2)	not determined			
			Ni_3Sn_4	Ni_3Sn_4	a=1218.4(5) b=405.3(2) c=521.7(2) β=105.97(3)°				
			Ni_3P	Fe_2P	a=586.5(1) c=338.59(1)				
			T5	Ni_2PSn	a=1282.40(2) b=359.411(6) c=508.98(1)				
NPS 85[*)]	$Ni_{30}P_{20}Sn_{50}$	200, 64d	(Sn)	βSn	a=583.186(2) b=318.231(2)	not determined			
			Ni_5P_4	Ni_5P_4	a=680.03(3) c=1102.67(8)				
			NiP_2	NiP_2	a=636.81(9) b=561.58(7) c=607.21(8) β=126.235(8)°				
			T5	Ni_2PSn	a=1281.44(3) b=359.170(8) c=508.82(1)				
NPS 86[*)]	$Ni_{30}P_{25}Sn_{45}$	200, 64d	(Sn)	βSn	a=583.200(2) b=318.239(2)	not determined			
			Ni_5P_4	Ni_5P_4	a=679.73(2) c=1103.09(7)				
			NiP_2	NiP_2	a=636.5(1) b=561.68(9) c=606.9(1) β=126.19(1)°				
			T5	Ni_2PSn	a=1280.94(7) b=359.15(2) c=508.85(3)				
NPS 87[*)]	$Ni_{30}P_{30}Sn_{40}$	200, 64d	(Sn)	βSn	a=583.14(4) b=318.13(3)	0.0	0.0	100.0	100.5
			Ni_5P_4	Ni_5P_4	a=679.82(2) c=1101.38(4)	55.4	44.1	0.5	100.3
			NiP_2	NiP_2	a=636.86(1) b=561.467(9) c=607.37(1) β=126.233(1)°	2.9	67.0	0.1	100.8
			T5	Ni_2PSn	a=1281.66(2) b=359.00(2) c=508.60(3)	9.8	24.6	25.6	101.1
NPS 88[*)]	$Ni_{30}P_{35}Sn_{35}$	200, 64d	(Sn)	βSn	a=583.158(4) b=318.230(3)	not determined			
			Ni_5P_4	Ni_5P_4	a=679.62(3) c=1101.35(8)				
			NiP_2	NiP_2	a=636.77(2) b=561.58(1) c=607.20(2) β=126.234(2)°				
			T5	Ni_2PSn	a=1281.80(5) b=359.08(1) c=508.85(2)				

Sample	Composition	Conditions	Phase label	Phase	Lattice parameters	Ni	P	Sn	Total
NPS 89[*]	$Ni_{30}P_{40}Sn_{30}$	200, 64d	(Sn) Ni_5P_4 NiP_2 T5	βSn Ni_5P_4 NiP_2 Ni_2PSn	$a=583.175(3)$ $b=318.242(2)$ $a=679.67(3)$ $c=1101.5(1)$ $a=636.76(1)$ $b=561.54(1)$ $c=607.22(1)$ $\beta=126.230(1)°$ $a=1281.89(6)$ $b=359.16(2)$ $c=508.69(3)$		not determined		
NPS 90[*]	$Ni_{15}P_2Sn_{83}$	200, 64d	(Sn) Ni_3Sn_4 T3	βSn Ni_3Sn_4 $Ni_{10}P_3Sn_5$	$a=583.176(6)$ $c=318.242(5)$ $a=1220.10(3)$ $b=405.98(1)$ $c=522.00(1)$ $\beta=105.19(2)°$ $a=647.4(3)$ $b=839.0(3)$ $c=1010.9(3)$ $\alpha=75.38(4)°$ $\beta=83.31(4)°$ $\gamma=84.02(3)°$	0.0 42.1 55.4	0.0 0.0 15.5	100.0 57.9 29.1	99.5 101.4 100.5
NPS 91[*]	$Ni_{30}P_2Sn_{68}$	200, 64d	(Sn) Ni_3Sn_4 T3	βSn Ni_3Sn_4 $Ni_{10}P_3Sn_5$	$a=583.204(3)$ $c=318.260(2)$ $a=1219.65(9)$ $b=405.88(3)$ $c=522.04(3)$ $\beta=105.229(5)°$ $a=646.3(7)$ $b=840.0(8)$ $c=1010.5(5)$ $\alpha=75.20(8)°$ $\beta=83.43(9)°$ $\gamma=84.00(5)°$	0.0 42.6 55.3	0.0 0.0 15.4	100.0 57.4 29.3	100.5 101.7 100.8

[*] sample stress annealed at 200°C before XRD
[c] sample composition corrected for P-loss
(l) = solidified from liquid during quenching
d = days
[Brno *] EDX measurements done at Inst. of Physics, CZ Academy of Sciences, Brno

T1 = $Ni_{10}P_3Sn$ T2 = $Ni_{21}P_6Sn_2$ T3 = $Ni_{10}P_3Sn_5$ T4 = $Ni_{13}P_3Sn_8$ T5 = Ni_2PSn

5. Results in the System Ni-P-Sn

This Chapter is based on Refs.[90] and [91]

Results of the phase analysis based on XRD and EPMA / EDX of samples annealed at 850, 700, 550 and 200 °C are shown in Table 5.1. According to the present results and to literature reports (Refs.[82-85]; see also Chapters 2.2 and 6) five ternary compounds and a large ternary solid solution based on the Ni_3Sn_2 HT-phase exist in the ternary Ni-P-Sn system; all of them are located in the Ni-rich part. Relevant crystallographic parameters and homogeneity ranges of these phases, as derived from the data in Table 5.1, are collected in Table 5.2.

Experiences gained during work on the Ni-P-Sn system showed that the determination of the P amount in the respective phases was problematic, because it was either over- or underestimated, depending on the use of EDX or WDS and also on the standard used for calibration. As this experimental problem could not be solved until now, for the construction of the ternary phase diagram the phase boundaries of unary and binary phases were taken from the respective binary phase diagrams whenever an insignificant amount of the respective third element (below 0.5 at.%) was measured.

Four partial isothermal sections are based on the results of the phase analysis (Table 5.1) and are shown in Figs. 5.1 – 5.3 and 5.16 (Chapter 5.6, page 98). The course of the liquid phase boundaries was derived from thermal analyses. Note that the apexes of three-phase fields representing the composition of the Sn-rich liquid could not unambiguously be determined due to decomposition of the liquid on cooling.

In these isotherms most phase fields could be derived from XRD and EPMA/EDX data, while the remaining ones were added consistently. A few samples appear to have slightly shifted from their nominal compositions (e.g. NPS 27, 35). Due to the narrow spacing of some of the phase fields these slight shifts caused the appearance of phases that are not in agreement with the nominal composition. Also note that at 850 and 700 °C phase fields that continue to the P- and Sn-rich parts are shown using dashed lines because these results are preliminary.

Before a description of the phase equilibria will be given, several aspects of the ternary phases with significance to the phase diagram shall be discussed.

Table 5.2: Space group, melting range and composition of ternary Ni-P-Sn Phases

Compound	Space Group	max. stability temp. [°C]	composition [at.%] this work	composition [at.%] literature
$Ni_{10}P_3Sn$ (T1)	$P3m1$	~1010 (congruent)	$Ni_{71.4}P_{21.4}Sn_{7.2}$	$Ni_{71.4}P_{21.4}Sn_{7.2}$ [82]
$Ni_{21}P_6Sn_2$ (T2)	$Fm\overline{3}m$	991 (congruent)	$Ni_{72.4}P_{20.7}Sn_{6.9}$	---
$Ni_{10}P_3Sn_5$ (T3)	$P\overline{1}$	~800	$Ni_{54.2}P_{16.6}Sn_{29.2} - Ni_{57.9}P_{15.3}Sn_{26.8}$ (550 °C) $Ni_{54}P_{16.7}Sn_{29.3} - Ni_{56.2}P_{15.8}Sn_{28.0}$ (700 °C)	$Ni_{55.6}P_{16.7}Sn_{27.7}$ (700 °C)[*)] [83]
$Ni_{13}P_3Sn_8$ (T4)	$P\overline{1}$	700 - 800	$Ni_{54.2}P_{11.8}Sn_{34} - Ni_{58.2}P_{11}Sn_{30.8}$ (550 °C) $Ni_{54.0}P_{13.5}Sn_{32.5}$ [**)] (700 °C)	$Ni_{54.2}P_{12.5}Sn_{33.3}$ (700 °C) [84]
Ni_2PSn (T5)	$Pnma$	722 (peritectic)	$Ni_{48.7}P_{25.3}Sn_{26} - Ni_{49}P_{24.5}Sn_{26.5}$	$Ni_{50}P_{25}Sn_{25}$ [85]
Ni_3Sn_2 HT	$P6_3/mmc$	---	limiting ternary comp.: $Ni_{52.9}P_{17.6}Sn_{29.5}$	maximum P-Conc.: 17 [36]

[*)] from chemical formula; no other composition given
[**)] from non-equilibrium samples

5.1 Ternary Ni-P-Sn phases

$Ni_{10}P_3Sn$ (T1)

T1 was described by Keimes and Mewis [82]. In the present work, both lattice parameter variation and composition according to EPMA / EDX suggest that this phase is a line compound. The stoichiometric composition $Ni_{71.4}P_{21.5}Sn_{7.1}$ is in agreement with the EPMA values, so that this phase has been placed at its stoichiometric position in the phase diagram (see Figs. 5.1 – 5.3). Its composition differs from neighbouring $Ni_{21}P_6Sn_2$ (T2) by only one atomic per cent in the Ni : P ratio. The structural relations between T1 and T2 will be highlighted in Chapter 6.

According to Keimes and Mewis [82], T1 has a melting point close to 850 °C. However, DTA results obtained from a sample with this stoichiometry (No. T1) show that there is no thermal effect anywhere close to this temperature, but that the first thermal effect to be observed by DTA occurs at 980 °C. It is rather weak, but can clearly be distinguished from the liquidus effect. This effect, however, cannot reflect a possible peritectic formation of T1, because in the neighboring sample NPS 51 a primary crystallization of this phase and a melting point higher than 980 °C were found which is not compatible with a peritectic reaction. It is therefore more likely that sample No. T1 was actually slightly off the stoichiometry and that one of the adjacent ternary reactions appeared in the DTA recording as well, most probably the reaction U12, L + T1 = Ni_3Sn_2 HT + T2 at 978 °C. The melting of the T1 phase was found at approx. 1010 °C.

$Ni_{21}P_6Sn_2$ (T2)

The compound T2 was found during work on the ternary phase diagram and its crystal structure will be described in Chapter 6. The structure is an ordered ternary variant of the C_6Cr_{23} structure type, which is common to many ternary borides and phosphides of the same stoichiometry. The present EPMA / EDX data scatter around the stoichiometry ($Ni_{72.4}P_{20.7}Sn_{6.9}$) of this compound, so that it was placed accordingly in the phase diagram. According to the lattice parameter it has been included as a line compound at the investigated temperatures. The result of a DTA measurement of a sample with the stoichiometry $Ni_{21}P_6Sn_2$ (No. T2) together with the fact that a single phase compound was obtained for powder XRD (see Chapter 6) suggest that the phase melts congruently at 991 °C.

Ternary Solid Solution of Ni_3Sn_2 HT

The ternary solid solution of Ni_3Sn_2 HT was characterized by EPMA and XRD at 850 °C. According to the present results this solid solution extends into the ternary up to 17.6 at.% P. This value is comparable with the P-rich limit of approx. 17 at.% determined by Furuseth and Fjellvåg [36] using the disappearing phase principle. In contrast to the present work these authors proposed a U-like shape for the homogeneity range based on an evaluation of the lattice parameters (see Fig. 2.4). However, they did not mention which of their samples were single- or multi-phase, respectively, and, according to the present results, most of their samples would have to be multi-phase samples, because they were not placed within the homogeneity range. In fact, they report that sample $Ni_{59.2}P_{12.2}Sn_{28.6}$ contained $Ni_{12}P_5$ LT in addition to Ni_3Sn_2 HT, which is consistent with the present evaluation. As a consequence, all other samples at higher Ni-contents should therefore not be single-phase either, for which no indication is given in the text of Ref. [36].

The lattice parameters of the binary Ni_3Sn_2 HT phase vary considerably from a=412.54, c=519.95 pm (Ni-rich side) to a=404.56, c=512.61 pm (Sn-rich limit) at 500 °C [42] (from Ref. [42] there are no values at 850 °C available. There is an increase of the lattice parameters with the temperature, but this variation is much smaller than the change with composition). In the ternary Ni-P-Sn system a very strong variation of the lattice parameters was noticed, too, at 850 °C: e.g. on the Ni-rich side a=403.61, c=516.37 pm corresponding to the composition $Ni_{57.8}P_{4.8}Sn_{37.4}$ (sample NPS 27) or a=387.16, c=521.10 pm corresponding to $Ni_{56.8}P_{14.5}Sn_{28.7}$ (NPS 28). The lattice parameters for the limiting composition at the corner of the three-phase field

[L + Ni_3Sn_2 HT + Ni_2P] were determined to be a=377.33, c=520.59 pm (NPS 29). It can be seen that the a-parameter constantly and strongly decreases, which is comparable with the evaluation of Furuseth and Fjellvåg [36].

In Ref. [86] Furuseth et al. reported the existence of complex modulated structures within the solid solution range around $Ni_{52.4}P_{14.3}Sn_{33.3}$, based on electron diffraction experiments, and proposed a structural model. NiAs-type structures are known for their flexibility and tendency to form superstructures, but as it is not clear if there is a true ternary compound at this composition, it has not been considered for the present phase diagram as an independent phase.

Between 850 and 700 °C the two ternary compounds $Ni_{10}P_3Sn_5$ (T3) [83] and $Ni_{13}P_3Sn_8$ (T4) [84] develop out of this large ternary solid solution, and the solubility of P in Ni_3Sn_2 HT was found to be reduced to 3.3 at.% at 550 °C.

$Ni_{10}P_3Sn_5$ (T3) and $Ni_{13}P_3Sn_8$ (T4)

Both, T3 [83] and T4 [84], were found by XRD and EPMA at 700 and 550 °C, and their homogeneity ranges at the corresponding temperatures are given in Table 5.2. Composition values for these two compounds can be derived from the crystal structure determinations mentioned in Refs. [83] and [84]: $Ni_{54.2}P_{12.5}Sn_{33.3}$ and $Ni_{55.6}P_{16.7}Sn_{27.7}$; they are in good agreement with the values obtained in the present work.

The crystal structures of both phases are reported to be triclinic super structures of the $InNi_2$-type structure of Ni_3Sn_2 HT [83, 84] and their formation out of the large ternary solid solution of Ni_3Sn_2 HT has been proposed. However, according to García-García et al. [83] the crystal structure of T3 was determined from a sample $Ni_{60}P_8Sn_{32}$ quenched from 700 °C containing the phases Ni_3Sn_2, Ni_2PSn (T5) and T3 (major component). In the present work no such three-phase field was found (see Fig. 5.2), and a sample at this nominal composition would in fact be placed within the three-phase field [Ni_3Sn_2 HT + $Ni_{12}P_5$ LT + T3].

Ni_2PSn (T5)

In the literature T5 was reported to have no appreciable homogeneity range around its stoichiometric composition [85]. In the present work EPMA data suggest a small homogeneity range of slightly less than 1 at.% in all directions at 550 °C. Furthermore, according to the present results this compound does not seem to be at its exact stoichiometric composition but slightly shifted towards the Sn-rich side.

At 850 °C T5 does not exist anymore and can occasionally be found in the matrix of partially liquid samples after quenching only. Therefore, the two-phase field [L + Ni_2P] is shown in Fig. 5.1 instead of the two-phase fields [Ni_2P + T5] and [L + T5]. This is basically in agreement with the literature, where the decomposition at 732 °C into Ni_2P and Sn (sic, should of course be the liquid) according to reaction $L + Ni_2P = T5$ has been proposed [85]. Such decomposition formally corresponds to a quasi-binary peritectic reaction. Indeed, in the present work suitable thermal effects at 722 °C were observed in the DTA measurements of samples placed on the section from Ni_2P to Sn, which fairly well agrees with the value of 732 °C in Ref. [85]. As these effects were only observed along this section but in no other samples, they support the findings of Furuseth and Fjellvåg [85].

Although quite rare, quasi-binary peritectic reactions have been described in the literature, e.g. in the Bi-Cd-In system (L + Bi = Z; Z denotes a ternary Bi-Cd-In phase) [92], in the Bi_2TeO_5-Bi_2SeO_5 system [93] or in the $CuGaTe_2$-HgTe system [94]. In Ni-P-Sn, the section from Ni_2P to Sn may even be a true quasi binary system, i.e. at all temperatures the relevant tie lines lie in the same vertical plane. T5 is located exactly on the direct connection of Ni_2P and Sn, and the relevant phase equilibria were found along these sections at the three investigated temperatures: $[L + Ni_2P]$ at 850 °C and [L + T5] as well as $[Ni_2P + T5]$ at 700 and 550 °C.

However, at 700 °C EDX data from sample NPS 30 suggest a widening of the homogeneity range towards the P-side. This would be inconsistent with the decomposition of T5 at 722 °C, where the phase width has to be reduced to a point. Furthermore, thermal effects slightly higher than 722 °C were observed in the neighbouring phase fields (DTA from samples 30 and 31), a fact which is incompatible with the development of the phase equilibria in this region. For a quasi-binary peritectic formation and the limited homogeneity range of T5 thermal effects on either side of the quasi-binary section would be expected to occur just below its temperature of 722 °C, where the resulting three-phase field $[L + Ni_2P + T5]$ would react with other phase fields. However, only thermal effects at higher temperature were observed (733 °C) that might contradict such a solution. Therefore, the suggested formation of T5 presented here should be regarded as tentative and further work will be necessary to unambiguously elucidate the phase relations in this area [95].

5.2 Phase Equilibria at 850 °C

Ni_3Sn_2 HT, $Ni_{12}P_5$ HT and Ni_2P are the highest congruently melting compounds in their respective binary systems. At 850 °C there is a huge ternary solid solution of P in Ni_3Sn_2 HT with a maximum P content of 17.6 at.% (see Chapter 5.1). The large two-phase fields [L + Ni_3Sn_2 HT] and [L + Ni_2P] can still be found at 850 °C and the liquid phase covers a comparatively large part of the Gibbs triangle (see Fig. 5.1a). The liquid phase frequently decomposes into several solid phases during cooling, which is nicely reflected by the results from samples NPS 36 and 37. In these samples up to three reaction products were formed out of the liquid during quenching. They could be found in the matrix of these samples (e.g. (Sn), P_3Sn_4 and T5 in sample NPS 37, compare Table 5.1).

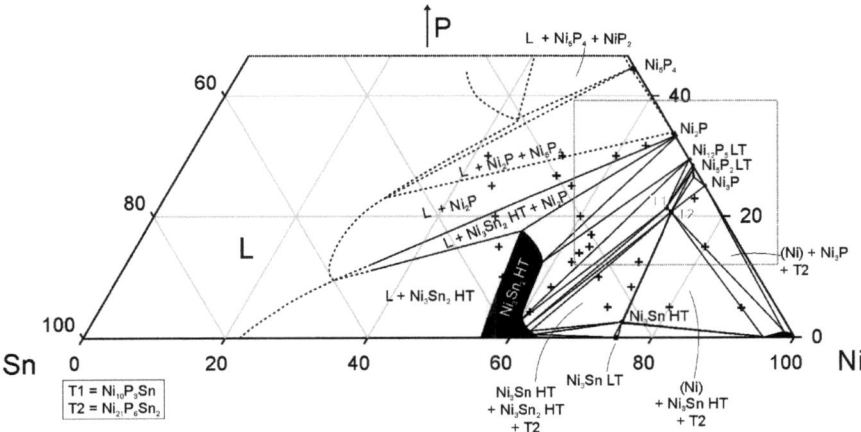

Fig. 5.1a: Partial isothermal section of the Ni-P-Sn system at 850 °C. '+' indicate nominal compositions of samples; for clarity not all sample positions are indicated in this Fig. (sample positions around T1 and T2 are shown in Fig. 5.1b). Uncertain phase equilibria and L-apexes of three-phase fields are shown by dashed lines. The grey dotted rectangle indicates the area shown in Fig. 5.1b.

On the Ni-rich side of Ni_3Sn_2 HT, broad two-phase fields connect Ni_3Sn_2 HT with $Ni_{12}P_5$ LT and Ni_2P, respectively. The solubility of Sn in the observed binary Ni-P phases generally is negligible. Only in the vicinity of Ni_5P_2 EPMA measurements revealed a phase with the composition $Ni_{72.9}P_{26.3}Sn_{0.8}$ (labelled "T", see Fig. 5.1b). This composition was found in sample NPS 79b ($Ni_{73}P_{25}Sn_2$; see also Table 5.1) in the as cast condition as well as after annealing at 550, 700 and 850 °C. Furthermore, in the diffraction patterns peaks were found that could not be indexed using structure data from any known binary or ternary phase. While in the ternary system a stabilization of Ni_5P_2 HT to lower temperatures is theoretically possible, it seems unlikely that this phase would even be found at 550 °C. In principle there could be a further ternary compound at composition T, or a solubility of Sn in Ni_5P_2 LT may extend until composition T. As there is currently no certainty for any interpretation, in the isothermal sections (Figs. 5.1-5.3) a dotted line indicates the extent of a possible ternary solid solution or the position of the ternary compound, respectively. Single crystal measurements are planned in order to clarify these uncertainties.

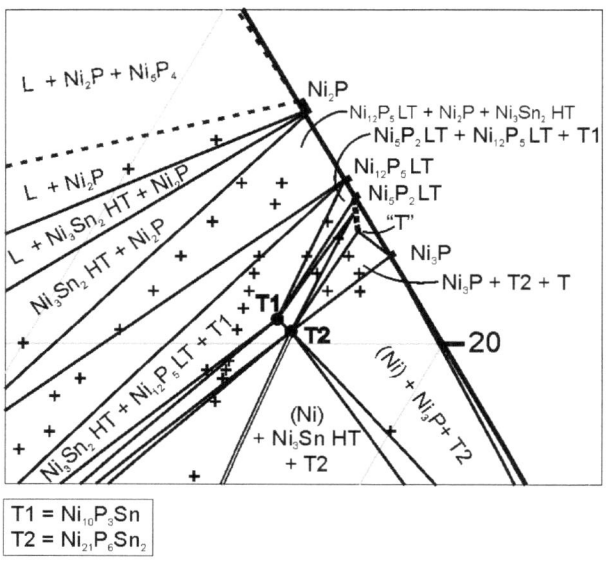

Fig. 5.1b: **Enlarged area around the compounds T1 and T2 of the isotherm at 850 °C. Symbols as in Fig. 5.1a.**

The phase equilibria shown around this composition T ($Ni_{72.9}P_{26.3}Sn_{0.8}$) are, of course, tentative. They comprise the three-phase fields [Ni_5P_2 LT + $Ni_{12}P_5$ LT + T1], [Ni_5P_2 LT + T1 + T2] and [Ni_3P + T2 + T]. This phase triangulation corresponds to the case where T is considered the limit of a ternary solid solution of Ni_5P_2 LT (as it is shown using dashed lines). If T was a true ternary compound, the related phase fields would have to be added accordingly.

Another prominent feature of the isotherm at 850 °C is the existence of the Ni_3Sn HT phase (BiF_3-, Fe_3Al-type, ordered DO_3 structure), which was found to dissolve 2.5 at.% P, resulting in the limiting composition of $Ni_{74.5}P_{2.5}Sn_{23}$. The appearance of this BiF_3 type phase at 850 °C and the fact that it can be retained by quenching are both remarkable, because the stabilization in the ternary

Ni-P-Sn system down to 850 °C is well below the binary transition temperatures of reactions p6 and e7 (948 and 911 °C, respectively). Furthermore, in the binary Ni-Sn system it cannot at all be obtained by quenching from temperatures around 1050 °C (see Ref. [42]). Thermal analysis showed that the HT – LT transformation occurs in a temperature range from 808 - 830 °C in the ternary system, so that in the isotherms at 700 and 550 °C the Ni_3Sn LT phase appears without any significant solubility of P (see also Chapter 5.4).

This behaviour does not seem to be uncommon with HT phases of the BiF_3 type structure. In the binary Cu-Sn system the γ-phase, Cu_3Sn HT, also has BiF_3 type structure and similarly to the situation of binary Ni_3Sn HT cannot be retained by quenching from high temperatures. In the ternary Cu-Ni-Sn system, where it forms a temperature dependent continuous solid solution with the isotypic Ni_3Sn HT phase, it cannot be obtained by quenching, either, as long as only a few at.% Ni are dissolved. However, around a Cu : Ni ratio of 1:1, not only quenching of this $(Cu,Ni)_3Sn$ phase is possible, but it is even stabilized to lower temperatures. It was experimentally found in samples quenched from 400 °C [40]. Thus, additions of Ni stabilize the Cu_3Sn γ-phase, and additions of Cu equally stabilize the Ni_3Sn HT phase. On the other hand, such stabilization was not found for Ni_3Sn HT in the Ag-Ni-Sn system, where it massively transforms into the hexagonal LT modification during quenching (Refs. [96] and [97]).

Considering the present results for Ni-P-Sn, it can be concluded that small elements may have a similar stabilizing effect as additions of other metals. While this stabilization was found for the addition of P, attempts to retain the Ni_3Sn HT phase by doping with 1 at.% of oxygen were not successful, as it is reported in Ref. [42]. It therefore can be resumed that the addition of certain amounts of third elements – other metals or small elements - can indeed stabilize BiF_3 type HT phases.

Thermodynamic and statistical thermodynamic studies for e.g. Fe_3Al and Ni_3Sb are available from the literature (Ipser et al. [98] and Huang et al. [99]). An expansion of these concepts for ternary solid solutions of these phases in combination with experimental results would certainly be interesting.

Phase equilibria on the P-rich side of the Ni_2P-Sn section are still rather tentative. Not only does the evaporation of P become significant at these compositions, but also annealing and quenching of the samples does not always yield equilibrium conditions. The existence of the metastable reaction $L = Ni_2P + NiP$ in the binary Ni-P system is likely to have a counterpart in the ternary system and may be the reason for the observed non-equilibrium. For instance, in sample NPS 38 which is placed in the three-phase field [$L + Ni_2P + Ni_5P_4$], five phases were found, among them NiP_2, which clearly points to non-equilibrium. The three-phase field [$L + Ni_2P + Ni_5P_4$] as shown in Figs. 5.1a and 5.1b was established according to results obtained from sample NPS 30 where T5 was found to be the result of decomposition of the liquid on cooling.

5.3 Phase Equilibria at 700 and 550 °C

The isothermal sections at 700 and 550 °C are shown in Figs. 5.2, 5.3a and 5.3b, respectively. In this Chapter, the Ni-rich phase equilibria will be discussed, while the Sn-rich phase equilibria will be the subject of Chapter 5.6.

At 700 °C (Fig. 5.2) the large homogeneity range of Ni_3Sn_2 HT has reduced to approx. 2 at.% P. The two ternary compounds T3 and T4 have formed out of this phase. The composition of T3 is close to the limiting P-concentration found in Ni_3Sn_2 HT at 850 °C (see Table 5.2). The formation mechanism of these two phases is not known at present, but thermal effects around 800 °C (797 °C according to Ref. [36]) appear to be related to this transition (see also Chapter 5.4).

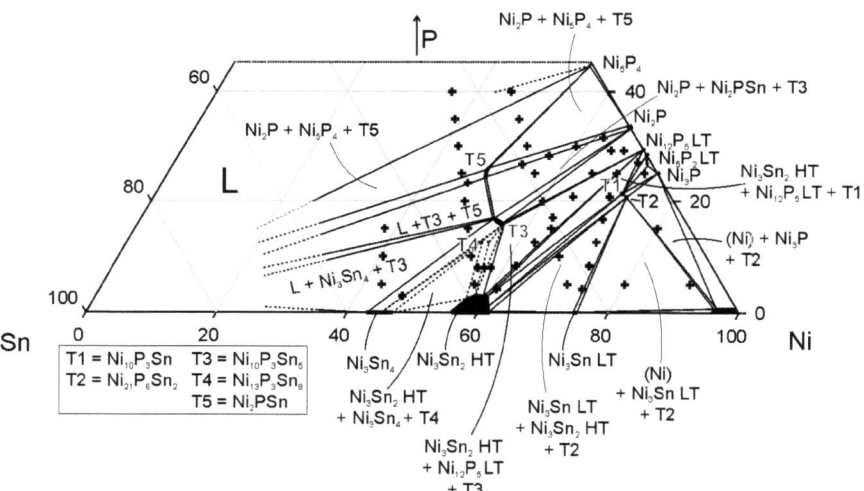

Fig. 5.2: Partial isothermal section of the Ni-P-Sn system at 700 °C. Symbols as in Fig. 5.1.

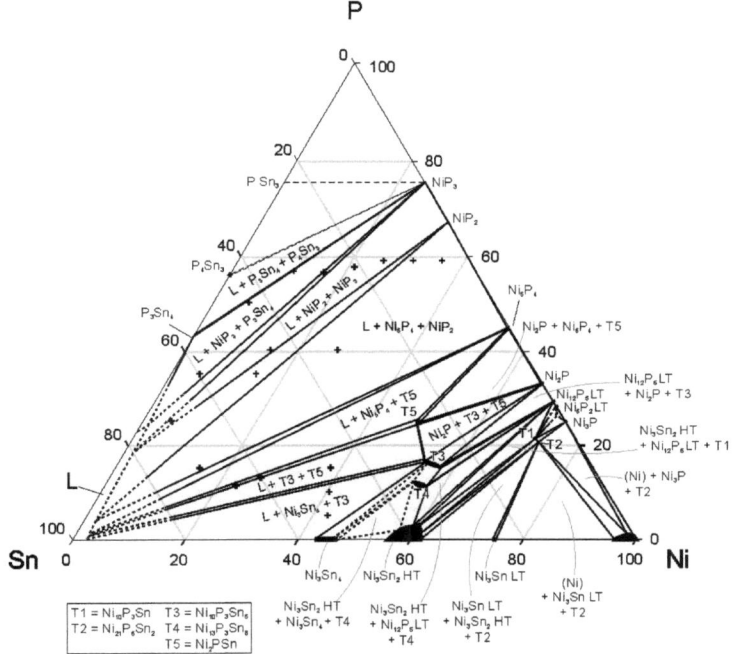

Fig. 5.3a: Isothermal section of the system Ni-P-Sn at 550 °C. Ambiguous phase equilibria or those added without experimental basis are shown using dashed lines. Symbols as in Fig. 5.1. An enlarged image of the Ni-rich part is shown in Fig. 5.3b.

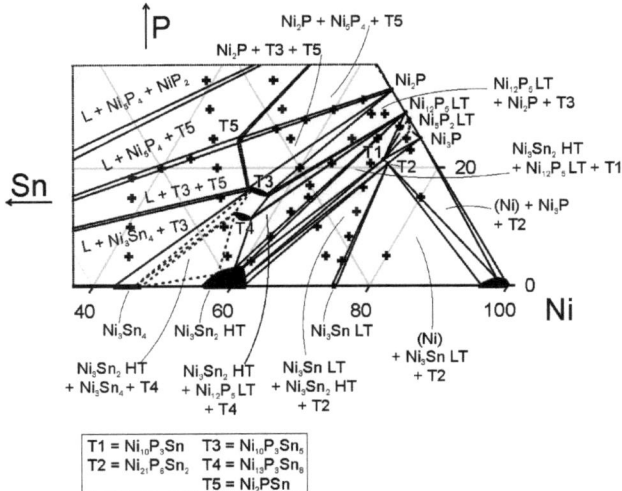

Fig. 5.3b: Enlarged image of the Ni-rich part of the isotherm at 550 °C. Symbols as in Fig. 5.1.

As outlined in Chapter 5.1, T5 is likely to be formed by a quasi-binary peritectic reaction at 722 °C and was therefore found in samples annealed at 700 and 550 °C (see Figs. 5.2, 5.3a and 5.3b) as an equilibrium phase. The formation of T3, T4 and T5 between 850 and 700 °C is accompanied by a significant change in the phase equilibria that requires the existence of several ternary invariant reactions between these two temperatures. As a result, the three-phase field [L + Ni_3Sn_2 HT + Ni_2P] (850 °C) has given way to the three-phase field [Ni_2P + T3 + T5] (700 °C) on the Ni-rich side of T3. Fig. 5.4 shows the microstructures (SEM images) of sample NPS 29 ($Ni_{56.25}P_{25}Sn_{18.75}$) placed in these phase fields after annealing and quenching from the respective temperatures.

Furthermore, at 850 °C there was a two-phase equilibrium [Ni_3Sn_2 HT + Ni_2P], while at 700 °C the phase fields [Ni_2P + T3] and [$Ni_{12}P_5$ LT + Ni_2P + T3] were observed instead. In addition, there is a large three-phase field [Ni_3Sn_2 HT + $Ni_{12}P_5$ LT + T3].

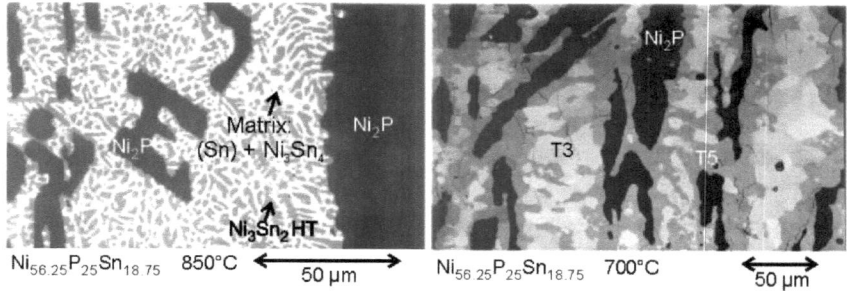

Fig. 5.4: SEM images of sample NPS 29 ($Ni_{56.25}P_{25}Sn_{18.75}$) annealed at 850 (left) and 700 °C (right). Different microstructures composed of different phases according to the three phase fields [L + Ni_3Sn_2 HT + Ni_2P] (850 °C) and [Ni_2P + T3 + T5] (700 °C) were obtained. On quenching from 850 °C the liquid decomposed into several phases, (Sn), Ni_3Sn_4 and T5. The amount of this latter phase was usually small and was in this sample only found by XRD.

Between 700 and 550 °C the phase equilibria in this area change once more, most likely according to the solid state Type II reaction U1, Ni_3Sn_2 HT + T3 = $Ni_{12}P_5$ LT + T4, in order to yield the three-phase fields [Ni_3Sn_2 HT + $Ni_{12}P_5$ LT + T4] and [$Ni_{12}P_5$ LT + T3 + T4] that were observed at 550 °C (Figs. 5.3a and 5.3b).

On the Sn-rich side of T3 the three-phase fields [L + T3 + T5] and [L + Ni_3Sn_4 + T3] were found according to the phase analysis. The remaining phase triangulation in the vicinity of Ni_3Sn_2 HT is shown by dotted lines, because samples prepared in this composition range and annealed at 700 °C did not allow an unambiguous clarification of the phase equilibria. Four phases, Ni_3Sn_2 HT, Ni_3Sn_4,

T3 and T4, were found in these clearly inhomogeneous samples (cf. samples NPS 100-102 in Table 5.1). Therefore, phase equilibria at 700 °C in this particular range are based on information from the surrounding phase triangulation and on powder XRD data from sample NPS 33, which revealed the two-phase field [Ni_3Sn_2 HT + T4].

Due to the same difficulties the extent of the homogeneity ranges of the phases T4 and Ni_3Sn_2 HT could not be determined unequivocally at 700 °C, either. Inhomogeneities, fine grained microstructures and precipitation of a dark phase within Ni_3Sn_2 HT grains hampered a reliable analysis by EPMA. In case of T4 the composition values derived from non-equilibrium samples, e.g. $Ni_{54.0}P_{13.5}Sn_{32.5}$ from sample NPS 34, are nevertheless in good agreement with the composition given by García-García et al. [84], $Ni_{54.2}P_{12.5}Sn_{33.3}$, and were subsequently used for the phase diagram construction. For Ni_3Sn_2 HT, EPMA values were found to scatter around 2 at.% P, which is comparable with the reports of Furuseth et al. [36] (3.3 at.% P at 557 °C).

Composition T was also found at 700 and 550 °C. There are slight changes in the phase triangulation in its vicinity. Instead of the three-phase field [Ni_3P + T + T2] at 850 °C, [Ni_3P + T + T1] was found at 550 and 700 °C. This change requires the existence of U-type invariant reaction between 850 and 700 °C: U3, T + T2 = Ni_3P + T1. However, no thermal effects pertinent to this reaction were found in the DTA measurements (Chapter 5.4).

The remainder of the Ni-rich phase equilibria shows two significant aspects, when compared to the situation at 850 °C. First of all, the Ni_3Sn LT-phase was found at 700 and 550 °C, which is consistent with the HT-LT transition taking place between 808 and 830 °C in the ternary system. In contrast to the HT modification, the LT phase does not show any appreciable ternary solubility of P. The second change involves the two-phase field [Ni_3Sn_2 HT + $Ni_{12}P_5$ LT]. This phase field appears to exist down to the temperature of the HT-LT transition in Ni_3Sn_2 (295 - 508 °C [42]) and can still be found at 550 °C. At this temperature, however, it has become significantly narrower.

On the P-rich side of the Ni_2P-Sn section the three-phase field [Ni_2P + Ni_5P_4 + T5] was found at both 700 and 550 °C based on sample NPS 30. For the remaining phase equilibria the same experimental difficulties as mentioned in Chapter 5.2 exist at the lower temperatures, too. Sample NPS 31 quite clearly defines the three-phase field [L + Ni_5P_4 + T5] at 700 °C. At 550 °C, however, samples NPS 37-40 suggest the existence of a three-phase field [L + NiP_2 + T5], while sample NPS 31 supports the existence of [L + Ni_5P_4 + T5] at 550 °C, too. These two three-phase fields cannot coexist at the same temperature. While the first version would be supported by a higher number of samples, there were no thermal effects between 550 and 700 °C that would allow a change of the phase equilibria. Therefore the three-phase field [L + Ni_5P_4 + T5] has tentatively been included in the isothermal section at 550 °C (Figs. 5.3a and 5.3b).

Table 5.3: Experimental results of the thermal analysis in the system Ni-P-Sn. All samples were measured in evacuated quartz crucibles at a heating rate of 5K/min.

No.	Nominal Comp. [at.%]	Heat Treatm. [°C]	Heating [°C] Invariant Effects	Heating [°C] Other Effects	Thermal Analysis Liquidus	Cooling [°C] Liquidus
Section Ni$_2$P - Sn						
NPS 3	Ni$_{63.33}$P$_{31.67}$Sn$_5$	550, 53d	731		1017	1009
NPS 4	Ni$_{60}$P$_{30}$Sn$_{10}$	550, 37d	727		999	996
NPS 5	Ni$_{56.67}$P$_{28.33}$Sn$_{15}$	700, 52d	722		953	946
NPS 6	Ni$_{53.33}$P$_{26.67}$Sn$_{20}$	700, 52d	723		933	915
Ni$_2$PSn	Ni$_{50}$P$_{25}$Sn$_{25}$	550, 43d	722		~910	905
NPS 7	Ni$_{46.67}$P$_{23.3}$Sn$_{30}$	700, 52d	722		905	896
NPS 8	Ni$_{43}$P$_{22}$Sn$_{35}$	700, 52d	231, 721		902	889
NPS 9	Ni$_{40}$P$_{20}$Sn$_{40}$	550, 48d	231, 720		888	883
NPS 10	Ni$_{37}$P$_{18}$Sn$_{45}$	550, 48d	231, 722		887	880
NPS 1	Ni$_{33.33}$P$_{16.67}$Sn$_{50}$	700, 7d	725		881	877
NPS 11	Ni$_{30}$P$_{15}$Sn$_{50}$	550, 48d	232, 722		877	874
NPS 12	Ni$_{27}$P$_{30}$Sn$_{60}$	550, 48d	231, 721		875	865
NPS 2	Ni$_{16.67}$P$_{8.33}$Sn$_{75}$	200, 125d	231, 723		867	845
NPS 13	Ni$_{23.33}$P$_{11.67}$Sn$_{65}$	550, 59d	232, 721		875	861
NPS 14	Ni$_{20}$P$_{10}$Sn$_{70}$	200, 125d	232, 726		858	856
NPS 15	Ni$_{13.33}$P$_{6.67}$Sn$_{80}$	200, 125d	231		>830	830
NPS 16	Ni$_{10}$P$_5$Sn$_{85}$	200, 125d	231		731	670
NPS 17	Ni$_{6.67}$P$_{3.33}$Sn$_{90}$	200, 125d	231		729	659
NPS 18	Ni$_{3.33}$P$_{1.67}$Sn$_{95}$	200, 125d	231		not observed	
Section Ni$_3$Sn - Ni$_3$P						
NPS 49	Ni$_{73.9}$P$_{4.2}$Sn$_{21.9}$	550, 169d	806, 945	1083	1132	1074
NPS 22	Ni$_{72.92}$P$_{8.3}$Sn$_{18.75}$	550, 169d	805, 949		1081	1029
NPS 50	Ni$_{71.9}$P$_{12.5}$Sn$_{15.6}$	550, 169d	947	1002	1068	989
NPS 23	Ni$_{70.8}$P$_{16.7}$Sn$_{12.5}$	550, 169d	944		983	923
NPS 104	Ni$_{70.5}$P$_{18}$Sn$_{11.5}$	850, 6d	978		994	936
NPS 105	Ni$_{70.4}$P$_{18.5}$Sn$_{11.1}$	850, 10d	978		993	926
NPS 106	Ni$_{70.3}$P$_{19}$Sn$_{10.7}$	850, 10d	976		994	941
NPS 51	Ni$_{69.8}$P$_{20.8}$Sn$_{9.4}$	550, 169d	941, 955	980	995	892

Section Ni₃Sn – Ni₂P (continued)						
NPS 107	Ni$_{69.5}$P$_{22}$Sn$_{8.5}$	850, 10d	941, 952	989	1000	
NPS 108	Ni$_{69.25}$P$_{23}$Sn$_{7.75}$	850, 10d	939, 952	979	1017	
NPS 109	Ni$_{69}$P$_{24}$Sn$_{7}$	850, 6d	940	958	1026	1007
NPS 24	Ni$_{68.75}$P$_{25}$Sn$_{6.25}$	550, 169d	950		1075	877
NPS 110	Ni$_{68}$P$_{28}$Sn$_{4}$	850, 10d	906, 982		1087	1069
NPS 52	Ni$_{67.7}$P$_{29.2}$Sn$_{3.1}$	550, 169d	911, 985	1020	1112	1062
NPS 111	Ni$_{67.3}$P$_{31}$Sn$_{1.7}$	850, 5d	906, 979		1103	1086
Section Ni₃Sn₂ - Ni₂P						
NPS 53	Ni$_{60.8}$P$_{4.2}$Sn$_{35}$	550, 169d		959		
NPS 19	Ni$_{61.7}$P$_{8.3}$Sn$_{30}$	550, 59d		940	1187	1149
NPS 54	Ni$_{62.5}$P$_{12.5}$Sn$_{25}$	550, 169d		931 – 949*)	1113	1089
NPS 115	Ni$_{62.8}$P$_{14}$Sn$_{23.2}$	850, 4d		923 – 948*)	1108	1083
NPS 20	Ni$_{63}$P$_{17}$Sn$_{20}$	550, 59d		917 – 945*)	1063	1038
NPS 114	Ni$_{63.7}$P$_{18}$Sn$_{18.3}$	850, 5d		921 – 948*)	1038	902
NPS 55	Ni$_{64.2}$P$_{20.8}$Sn$_{15}$	550, 169d	910, ~930		940	870
NPS 113	Ni$_{64.7}$P$_{23}$Sn$_{12.3}$	850, 4d	906	931	957	926
NPS 21	Ni$_{65}$P$_{25}$Sn$_{10}$	550, 59d	800, 904		979	924
NPS 112	Ni$_{65.5}$P$_{27}$Sn$_{7.5}$	850, 5d	907	984 max		
NPS 56	Ni$_{65.8}$P$_{29.2}$Sn$_{5}$	550, 169d	909	986 max	1034	1028
Section Ni₃Sn₂ - P						
NPS 25	Ni$_{71.25}$P$_{5}$Sn$_{23.75}$	550, 19d	812, 953		1131	1117
NPS 26	Ni$_{67.5}$P$_{10}$Sn$_{22.5}$	550, 19d	950		980	896
NPS 27	Ni$_{63.75}$P$_{15}$Sn$_{21.25}$	550, 19d		960	1073	1027
NPS 28	Ni$_{60}$P$_{20}$Sn$_{20}$	550, 19d	912		1005	864
NPS 29	Ni$_{56.25}$P$_{25}$Sn$_{18.75}$	550, 19d	735, 801	879	950	941
NPS 30	Ni$_{52.5}$P$_{30}$Sn$_{17.5}$	550, 19d	725	767	935	928
Section Ni₃Sn₂ - P						
NPS 33	Ni$_{57}$P$_{5}$Sn$_{38}$	550, 59d			957	943
NPS 34	Ni$_{54}$P$_{10}$Sn$_{36}$	550, 19d		792, 963	1126	1093
NPS 35	Ni$_{51}$P$_{15}$Sn$_{34}$	550, 19d		831	916	841
NPS 36	Ni$_{48}$P$_{20}$Sn$_{32}$	550, 19d	740	800 – 805, 856	> 892	892

No.	Nominal Comp. [at.%]	Heat Treatm. [°C]	Thermal Analysis			
			Invariant Effects	Heating [°C] Other Effects	Liquidus	Cooling [°C] Liquidus
Section $Ni_3Sn_2 - P$ (continued)						
NPS 37	$Ni_{45}P_{25}Sn_{30}$	550, 19d	736		897	894
NPS 38	$Ni_{42}P_{30}Sn_{28}$	550, 19d	731	629, 772	857	843
NPS 39	$Ni_{39}P_{35}Sn_{26}$	550, 19d		623	789	769
NPS 40	$Ni_{36}P_{40}Sn_{24}$	550, 19d		623	784	738
Section $Ni_3Sn_4 - P$						
NPS 41	$Ni_{42.75}P_5Sn_{52.25}$	550, 28d	782	888	1008	963
NPS 42	$Ni_{40.5}P_{10}Sn_{49.5}$	550, 28d		811 – 822	873	860
NPS 43	$Ni_{38.25}P_{15}Sn_{46.75}$	550, 28d	733 – 744	850	889	866
NPS 48	$Ni_{27}P_{40}Sn_{33}$	550, 28d		748	865	864
Section at 5 at.% Ni						
NPS 66	$Ni_5P_5Sn_{90}$	200, 102d	232			660
NPS 67	$Ni_5P_{10}Sn_{85}$	200, 102d	230		685	657
NPS 68	$Ni_5P_{15}Sn_{80}$	200, 102d	232	488	683	650
NPS 69	$Ni_5P_{20}Sn_{75}$	200, 102d	232		526	509
NPS 70	$Ni_5P_{25}Sn_{70}$	200, 102d	231	543	755	706
NPS 71	$Ni_5P_{30}Sn_{65}$	200, 102d	231	557, 663	751	697
NPS 72	$Ni_5P_{35}Sn_{60}$	200, 102d	232	558, 667	739	696
NPS 73	$Ni_5P_{40}Sn_{55}$	200, 102d	231	~535 – 562	769	727
Section at 15 at.% Ni						
NPS 90	$Ni_{15}P_2Sn_{83}$	200, 64d	232		831	780
NPS 74	$Ni_{15}P_5Sn_{80}$	200, 102d	231, 723		827	822
NPS 75	$Ni_{15}P_{15}Sn_{70}$	200, 102d	232	729	780	762
NPS 76	$Ni_{15}P_{25}Sn_{60}$	200, 102d	230	~710	749	719
NPS 77	$Ni_{15}P_{35}Sn_{50}$	200, 102d	232	524	767	690
NPS 78	$Ni_{15}P_{40}Sn_{45}$	200, 102d	230	543, 820	840	769

Section at 30 at.% Ni				
NPS 91	$Ni_{30}P_2Sn_{68}$	200, 64d	232, 783	512, 976	1059	1029
NPS 82	$Ni_{30}P_5Sn_{65}$	200, 64d	231	761	854	809
NPS 83	$Ni_{30}P_{10}Sn_{60}$	200, 64d	231	814	870	859
NPS 84	$Ni_{30}P_{15}Sn_{55}$	200, 64d	232, 729		875	867
NPS 85	$Ni_{30}P_{20}Sn_{50}$	200, 64d	233		851	839
NPS 86	$Ni_{30}P_{25}Sn_{45}$	200, 64d	232		771	732
NPS 87	$Ni_{30}P_{30}Sn_{40}$	200, 64d	233		776	750
NPS 88	$Ni_{30}P_{35}Sn_{35}$	200, 64d	233	752	776	742
NPS 89	$Ni_{30}P_{40}Sn_{30}$	200, 64d	233	770	712	764

Additional Samples				
T1	$Ni_{71.43}P_{21.43}Sn_{7.14}$	700, 24d	981		1010	967
T2	$Ni_{72.41}P_{20.69}Sn_{6.9}$	700, 21d			991	953
NPS 62	$Ni_{74.5}P_{23}Sn_{2.5}$	550, 59d	861, 951	971	997	978
NPS 63	$Ni_{90}P_5Sn_5$	550, 59d	859, 872		not observed	
NPS 64	$Ni_{80}P_5Sn_{15}$	550, 59d	~830, 872	1060	1087	1076
NPS 65	$Ni_{90}P_{15}Sn_5$	550, 59d	861, 874			
NPS 79	$Ni_{71}P_{27}Sn_2$	700, 21d	960		1129	1110
NPS 79b	$Ni_{73}P_{25}Sn_2$	850, 21d	945, 964	1015	1065	1048
NPS 116	$Ni_{72}P_{26}Sn_2$	850, 8d	985	1004, 1046	1100	1067
NPS 117	$Ni_{74}P_{24}Sn_2$	850, 8d	947	970, 1013	1032	1022
NPS 118	$Ni_{71}P_{25}Sn_4$	850, 8d	962	1003	1081	1056
NPS 119	$Ni_{72}P_{24}Sn_4$	850, 8d	986		1051	1013
NPS 120	$Ni_{73}P_{23}Sn_4$	850, 8d	947	991	1012	997

*) sequence of overlapping peaks

Fig. 5.5: Scheil Diagram of the Ni-rich area of the Ni-P-Sn phase diagram. T1 = $Ni_{10}P_3Sn$, T2 = $Ni_{21}P_6Sn_2$

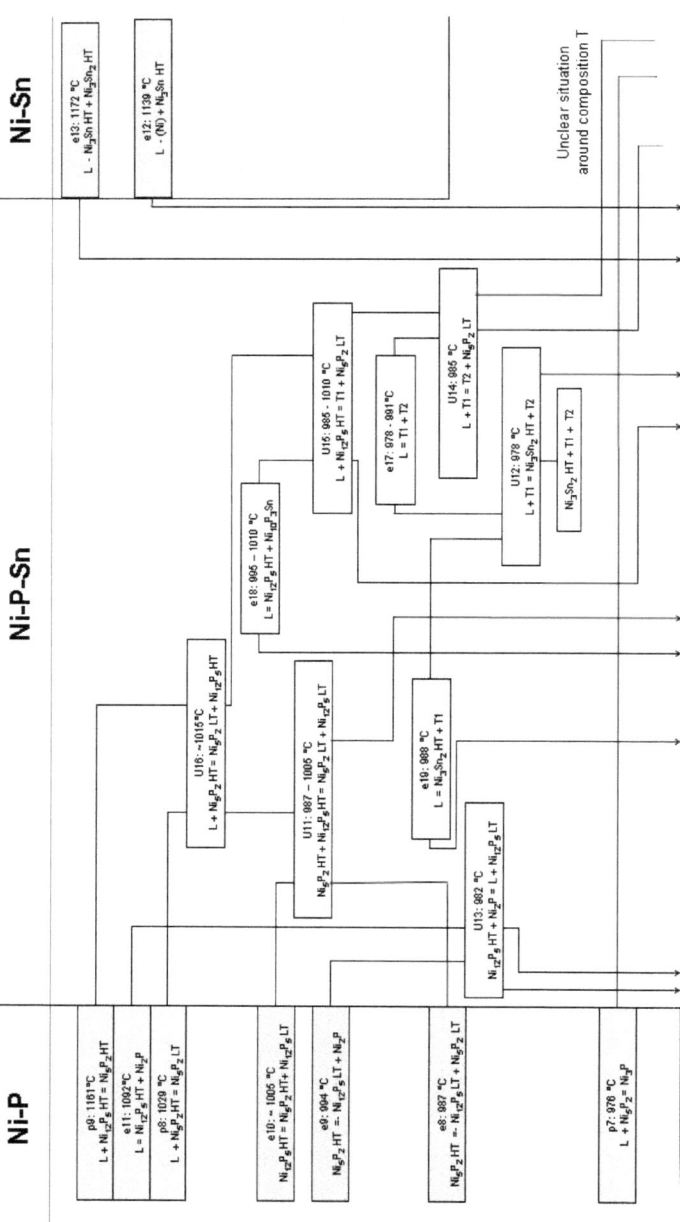

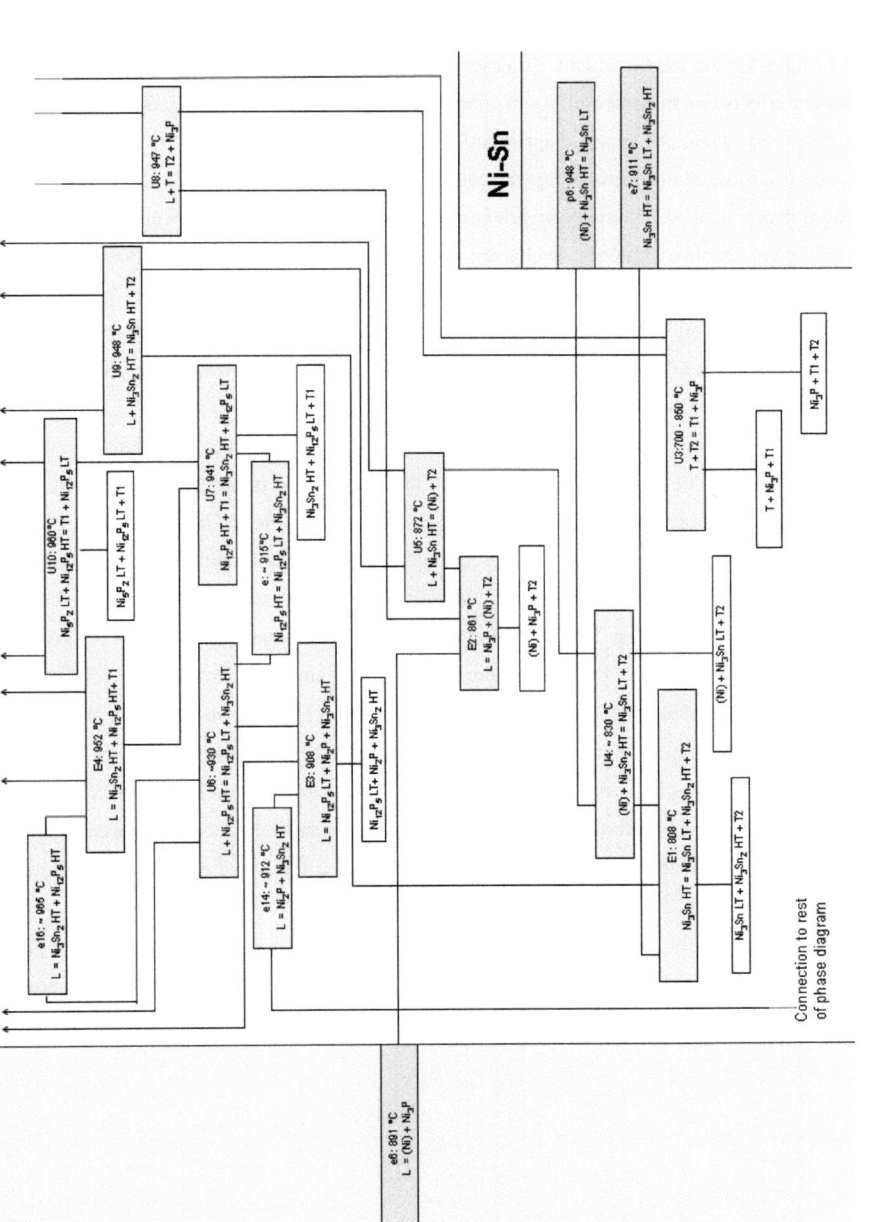

5.4 Thermal Behaviour in the Ni-rich part

The results of the thermal analyses (DTA), carried out on samples annealed at 200, 550, 700 or 850 °C, are compiled in Table 5.3. The listed temperatures represent the average values of first and second heating for the measurements. In case of a larger scatter for the liquidus temperature or monovariant effects due to non-equilibrium conditions after the first DTA cycle, the value from the first heating was taken. All phase diagram information relies on temperatures from the heating curves, because substantial supercooling was noticed in the recorded cooling curves.

From these data a total of 27 ternary invariant reactions in the Ni-rich corner were determined, of which five are quasi binary eutetics, twelve are true ternary reactions that involve the liquid phase, eight are solid state reactions and two are the melting points of T1 ($Ni_{10}P_3Sn$) and T2 ($Ni_{21}P_6Sn_2$). They are summarized in Table 5.4 with their reaction equation, temperature and type. The liquid composition was deduced from the DTA measurements in consistency with other phase diagram information (Chapters 5.1-5.3).

Information from the DTA was further enhanced by metallographic investigations of the solidification behavior of a number of samples with the focus on the primary crystallization. These results are shown in Table 5.5 where the primary crystallization is indicated by an asterisk.

A detailed overview over the nature and sequence of ternary invariant reactions (four-phase equilibria) and the connections to the binary boundary systems is given by the Scheil Diagram (reaction scheme) shown in Fig. 5.5. The composition of the participating liquid phase is given by the partial liquidus projection in Fig. 5.6., where the connecting monovariant lines (liquidus valleys) and the primary crystallization fields can also be seen. Three (partial) vertical sections, Ni_3Sn - Ni_2P, Ni_3Sn - P and Ni_3Sn_2 - Ni_2P, were derived from all experimental data (Figs. 5.9-5.12). The section from Ni_3Sn – P is only shown to a P content of 35 at.% due to increased experimental difficulties at higher P concentrations. The phase field boundaries in these isopleths were drawn in consistency with the invariant reactions listed in Table 5.4 and the phase equilibria shown in the isothermal sections (Figs. 5.1 to 5.3).

It is remarkable that in the Ni-rich corner most of the invariant reactions involving the liquid phase occur between ~1015 and 861 °C, i.e. in a narrow temperature interval of approx. 150 °C. Therefore they are at higher temperatures than the highest investigated isothermal section (850 °C) (Fig. 5.1).

Table 5.4: Invariant reactions in the system Ni-P-Sn

Reaction	Designation in this work	Type	Temperature [°C]
Ni_3Sn_2 HT + T3 = $Ni_{12}P_5$ LT + T4	U1	solid state reaction	550 - 700
T + T2 = Ni_3P + T1	U3	solid state reaction	700 - 850
Ni_3Sn HT = Ni_3Sn_2 HT + Ni_3Sn LT +T2	E1	ternary eutectoid	808
(Ni) + Ni_3Sn HT = Ni_3Sn LT +T2	U4	solid state reaction	~ 830
L = (Ni) + Ni_3P + T2	E2	ternary eutectic	861
L + Ni_3Sn HT = (Ni) + T2	U5	transition reaction	872
L = $Ni_{12}P_5$ LT + Ni_2P + Ni_3Sn_2 HT	E3	ternary eutectic	908
L = Ni_3Sn_2 HT + Ni_2P	e14	quasi binary eutectic	912
$Ni_{12}P_5$ HT = Ni_3Sn_2 HT + $Ni_{12}P_5$ LT	e15	quasi binary eutectoid	~ 915
L + $Ni_{12}P_5$ HT = Ni_3Sn_2 HT + $Ni_{12}P_5$ LT	U6	transition reaction	~930
$Ni_{12}P_5$ HT + T1 = Ni_3Sn_2 HT + $Ni_{12}P_5$ LT	U7	solid state reaction	941
L + T = Ni_3P + T2	U8	transition reaction	947
L + Ni_3Sn_2 HT = Ni_3Sn HT + T2	U9	transition reaction	948
L = Ni_3Sn_2 HT + $Ni_{12}P_5$ HT + T1	E4	ternary eutectic	952
Ni_5P_2 LT + $Ni_{12}P_5$ HT = T1 + $Ni_{12}P_5$ LT	U10	solid state reaction	960
Ni_5P_2 HT + $Ni_{12}P_5$ HT = Ni_5P_2 LT + $Ni_{12}P_5$ LT	U11	solid state reaction	987 - 1005
L = Ni_3Sn_2 HT + $Ni_{12}P_5$ HT	e16	quasi binary eutectic	~ 965
L + T1 = Ni_3Sn_2 HT + T2	U12	transition reaction	978
L = T1 + T2	e17	quasi binary eutectic	978 - 991
$Ni_{12}P_5$ HT + Ni_2P = L + $Ni_{12}P_5$ LT	U13	transition reaction	982
L + T1 = Ni_5P_2 LT + T2	U14	transition reaction	985
L + $Ni_{12}P_5$ HT = T1 + Ni_5P_2 LT	U15	transition reaction	985 - 1010
L = $Ni_{12}P_5$ HT + T1	e18	quasi binary eutectic	995 - 1005
L = Ni_3Sn_2 HT + T1	e19	quasi binary eutectic	988
L = T2	melting	congruent	991
L = T1	melting	congruent	1010
L + Ni_5P_2 HT = Ni_5P_2 LT + $Ni_{12}P_5$ HT	U16	transition reaction	~1015

Table 5.5: Results of the analysis of as-cast samples; all samples were air cooled from 1180 °C. '*' denotes the primary crystallization.

No.	Nominal Composition [at.%]	Phase	Primary Crystallization	Lattice Param. [pm]	EDX / WDS Ni[at.%]	P[at.%]	Sn[at.%]
NPS 20	$Ni_{63}P_{17}Sn_{20}$	Ni_3Sn_2 HT	*	a=399.15(2) c=517.42(3)	58.3	37.6	4.1
		$Ni_{12}P_5$ LT		a=864.91(4) c=507.53(3)	69.1	27.8	3.1
NPS 21	$Ni_{65}P_{25}Sn_{10}$	Ni_3Sn_2 HT		a=390.85(2) c=519.44(3)			
		$Ni_{12}P_5$ LT	* or	a=864.49(3) c=507.26(2)	70.6	29.2	0.2
		Ni_2P	*	a=586.35(3) c=339.05(3)	66.7	33.1	0.2
NPS 23	$Ni_{70.8}P_{16.7}Sn_{12.5}$	Ni_3Sn_2 HT	*	a=413.78(6) c=520.88(1)	62.6	1.8	35.6
		$Ni_{12}P_6Sn_2$		a=1111.55(1)	72.1	20.3	7.5
NPS 24	$Ni_{68.75}P_{25}Sn_{6.25}$	Ni_3Sn_2 HT	*	a=406.752(7) c=516.83(1)	70.5	28.5	1.0
		$Ni_{12}P_5$ LT		a=864.76(1) c=507.543(7)			
NPS 25	$Ni_{71.25}P_5Sn_{23.75}$	$Ni_{21}P_6Sn_2$		a=1111.61(8)	72.1	19.4	8.5
		Ni_3Sn_2 HT	*	a=414.170(7) c=521.34(1)	61.6	0.9	37.5
		Ni_3Sn HT		a=586.36(3)	73.5	2.3	24.2
		unknown			71.6	17.4	11.0
NPS 26	$Ni_{67.5}P_{10}Sn_{22.5}$	$Ni_{21}P_6Sn_2$		a=1111.80(2)			
		Ni_3Sn_2 HT	*	a=414.936(7) c=521.17(1)			
NPS 27	$Ni_{63.75}P_{15}Sn_{21.25}$	Ni_3Sn_2 HT	*	a=404.375(9) c=516.49(2)	58.5	2.9	38.7
		$Ni_{12}P_5$ LT		a=864.71(3) c=507.39(2)			
NPS 28	$Ni_{60}P_{20}Sn_{20}$	Ni_2P		a=586.48(4) c=338.98(4)			
		Ni_3Sn_2 HT	*	a=387.37(1) c=520.88(2)	56.3	14.3	29.5
NPS 29	$Ni_{56.25}P_{25}Sn_{18.75}$	Ni_2P	*	a=372.83(1) c=518.34(4)	65.7	34.2	0.1
		Ni_3Sn_2 HT		a=586.17(3) c=338.54(2)	51.7	18.0	30.3
		(Sn)		a=583.03(2) c=318.14(1)	not determined		
NPS 50	$Ni_{71.9}P_{12.5}Sn_{15.6}$	$Ni_{21}P_6Sn_2$		a=1112.93(6)	72.3	20.2	7.5
		Ni_3Sn_2 HT	*	a=415.08(5) c=520.71(6)	62.8	1.4	35.8
		Ni_3Sn HT		a=584.76(4)	72.1	2.7	25.2

No.	Nominal Composition [at %]	Phase	Primary Crystallization	Lattice Param. [pm]	EDX / WDS Ni[at.%]	P[at.%]	Sn[at.%]
NPS 51	$Ni_{69.8}P_{20.8}Sn_{9.4}$	$Ni_{10}P_3Sn$	*	a=766.38(1) c=963.01(2)	71.2	21.7	7.1
		$Ni_{12}P_5$ LT		a=864.82(4) c=507.41(3)	70.9	26.8	2.3
		Ni_3Sn_2 HT		a=405.42(1) c=516.45(2)	59.8	3.6	36.6
NPS 52	$Ni_{67.7}P_{29.2}Sn_{3.1}$	$Ni_{12}P_5$ LT	* or	a=864.55(1) c=507.203(7)	70.9	29.0	0.1
		Ni_2P	*	a=586.56(2) c=339.14(2)	67.1	32.7	0.2
		Ni_3Sn_2 HT		a=386.75(4) c=520.07(8)	in matrix; too fine for measurement		
NPS 53	$Ni_{60.8}P_{4.2}Sn_{35}$	$Ni_{12}P_5$ LT	*	a=864.5(2) c=507.9(2)			
		Ni_3Sn_2 HT		a=409.010(6) c=517.572(9)			
NPS 54	$Ni_{62.5}P_{12.5}Sn_{25}$	$Ni_{12}P_5$ LT	*	a=865.06(6) c=507.48(4)			
		Ni_3Sn_2 HT		a=403.91(2) c=516.41(4)			
NPS 55	$Ni_{64.2}P_{20.8}Sn_{15}$	$Ni_{12}P_5$ LT	no prim. Cryst	a=864.48(3) c=507.23(2)			
		Ni_3Sn_2 HT	discernible	a=389.94(1) c=519.77(2)			
NPS 56	$Ni_{65.8}P_{29.2}Sn_5$	Ni_2P	*	a=586.34(2) c=338.89(2)	66.9	32.9	0.2
		$Ni_{12}P_5$ LT		a=873.8(2) c=503.0(2)	70.8	29.1	0.2
		Ni_3Sn_2 HT		a=388.24(3) c=520.93(6)	too fine for measurement		
NPS 79	$Ni_{71}P_{27}Sn_2$	Ni_5P_2 LT	* or	a=662.31(4) c=1225.2(1)	70.9	28.1	1.0
		$Ni_{12}P_5$ LT	*	a=864.93(4) c=507.58(3)	70.1	29.1	0.8
		$Ni_{10}P_3Sn$		a=766.43(6) c=964.0(1)	70.5	22.3	7.2
		Ni_3Sn_2 HT		a=407.92(9) c=519.4(2)	69.0	19.1	11.9
NPS 79b	$Ni_{73}P_{25}Sn_2$	$Ni_{21}P_6Sn_2$	*	a=1109.2(2)	72.3	20.9	6.8
		T		---	72.6	26.3	1.2
		Ni_3P		a=880.4(3) c=441.7(2)	74.5	24.5	0.9
T1	$Ni_{71.43}P_{21.43}Sn_{7.14}$	$Ni_{10}P_3Sn$	*	a=767.23(1) c=962.74(2)	71.1	21.4	7.4
		$Ni_{12}P_5$ LT		a=871.6(6) c=497.4(4) very low amount	no individual measurements		
		Ni_3Sn_2 HT		a=408.58(8) c=519.1(2) very low amount			
		matrix			68.2	19.1	12.7
		matrix			70.0	17.2	12.8
T2	$Ni_{72.41}P_{20.69}Sn_{6.9}$	$Ni_{21}P_6Sn_2$	*	a=1111.73(1)	71.0	21.8	7.2
		matrix			68.9	19.9	11.2

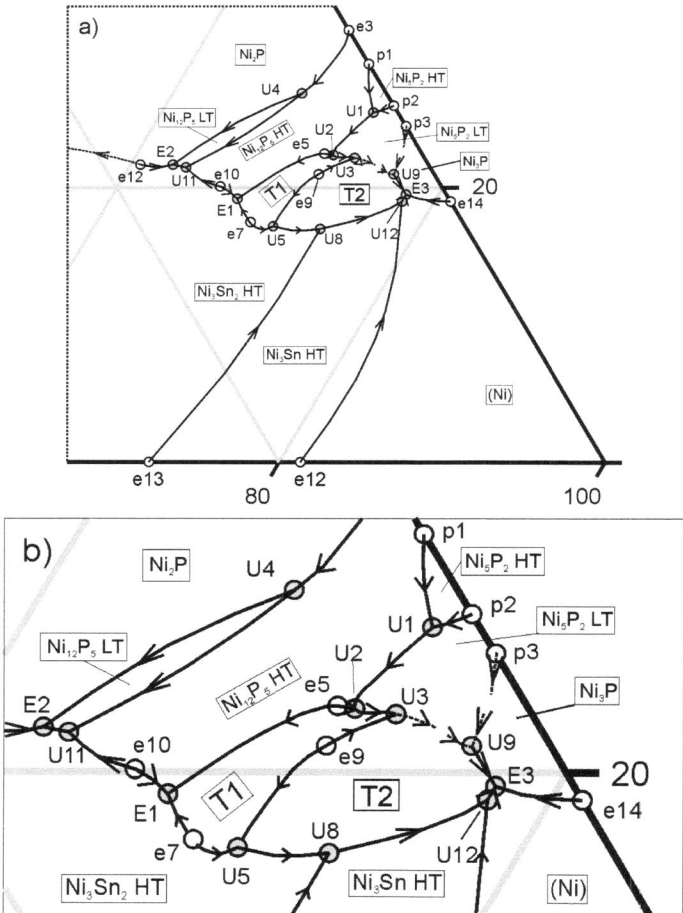

Fig. 5.6: a) Liquidus projection of the Ni-rich part of the system Ni-P-Sn showing binary and ternary invariant reactions and their connections via liquidus valleys. The primary crystallization is indicated by the framed text. b) Enlarged part around the primary crystallization fields of T1 and T2.

The primary crystallization fields of the binary compounds with the highest congruent melting points, Ni_3Sn HT, Ni_3Sn_2 HT, $Ni_{12}P_5$ HT and Ni_2P, were found to be quite large (cf. Fig. 5.6). A quasi-binary eutectic reaction (e14) exists between Ni_2P and Ni_3Sn_2 HT: L = Ni_2P + Ni_3Sn_2 HT at 912 °C (identical to a critical tie-line). The resulting microstructure can be seen in the micrograph of as-cast sample NPS 28 in Fig. 5.7 as an extremely fine eutectic microstructure comprising two phases: Ni_3Sn_2 HT + Ni_2P. For $Ni_{12}P_5$ HT and Ni_3Sn_2 HT a similar quasi binary eutectic is proposed here, based on the fine microstructure observed in as-cast sample NPS 24 (Fig. 5.8), i.e.

e16: L = $Ni_{12}P_5$ HT + Ni_3Sn_2 HT at approx. 965 °C. The reaction temperature is based on the course of monovariant effects obtained from DTA curves (shown in the isopleth in Fig. 5.12) that suggest a maximum above 950 °C. Further DTA effects suggest a minimum at about 915 °C in the three-phase field [$Ni_{12}P_5$ HT + Ni_3Sn_2 HT + $Ni_{12}P_5$ LT], which was interpreted as the quasi-binary eutectoid decomposition of the $Ni_{12}P_5$ HT phase, i.e. e15: $Ni_{12}P_5$ HT = Ni_3Sn_2 HT + $Ni_{12}P_5$ LT. Both reaction temperatures, however, are well below the binary transition temperatures of $Ni_{12}P_5$ LT and HT (e10, ~1005 and e9, 994 °C). As a consequence, the $Ni_{12}P_5$ HT phase would need to be stabilized to lower temperatures in the ternary system, where it is assumed to dissolve a small amount of Sn and to protrude over the LT phase. Of course, this result is somewhat tentative and can hardly be proven because the $Ni_{12}P_5$ HT phase cannot be stabilized by quenching in the ternary system, either. Nevertheless, the situation proposed here is fully consistent with the development of the phase equilibria in this area (see below). The internal structure of what appears to have been the primary $Ni_{12}P_5$ HT grains in Fig. 5.8 may be explained by the eutectoid decomposition of this phase. This effect would be very pronounced in sample NPS 24, because it is situated closest to the eutectic and eutectoid reactions.

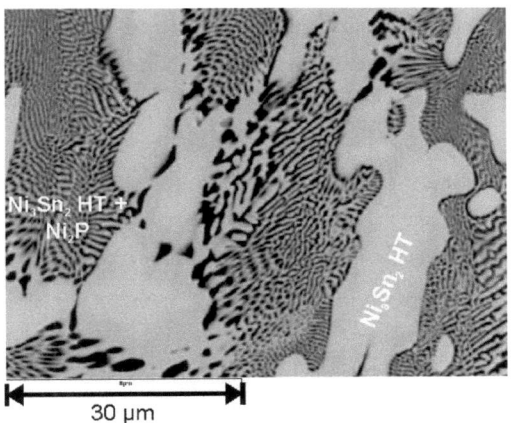

Fig. 5.7: SEM image of as-cast sample NPS 28. The primary crystallization of Ni_3Sn_2 HT and the eutectic matrix of Ni_3Sn_2 HT + Ni_2P can be seen.

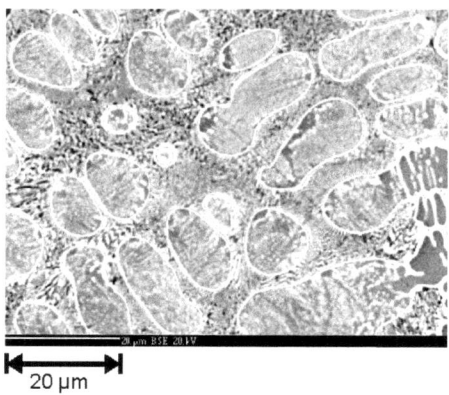

Fig. 5.8: SEM image of as-cast sample NPS 24. The primary crystallization of Ni_3Sn_2 HT and the eutectic matrix of Ni_3Sn_2 HT + $Ni_{12}P_5$ LT can be seen.

Note that none of the quasi binary reactions described so far are part of a true quasi binary system. The compositions of the involved solid phases do not always represent congruent melting points (e.g. in Ni_3Sn_2 HT). At temperatures different from the reaction temperatures the tie-lines as well as e.g. reactions e15 and e16 do not necessarily lie in the same vertical plane, either. The eutectic points of e14 and e16 are maxima in the respective monovariant liquidus valleys (and thus in the related three-phase fields), but are in fact saddle points in the liquidus surface (Figs. 5.6, 5.9-5.12). The e-type reactions occur due to the joining of two two-phase fields comprising the liquid and one of the solid compounds each, and thus are in fact critical tie-lines.

At temperatures below the reactions e14 to e16 the two-phase fields [Ni_3Sn_2 HT + Ni_2P] and [Ni_3Sn_2 HT + $Ni_{12}P_5$ HT] or [Ni_3Sn_2 HT + $Ni_{12}P_5$ LT], respectively, exist. At 850 °C these phase fields were found to be very broad (see Chapter 5.2).

In the binary Ni-P system there is a eutectic reaction e11, L = $Ni_{12}P_5$ HT + Ni_2P (1092 °C). From this binary eutectic the corresponding three phase field opens into the ternary. Together with the three-phase field [$Ni_{12}P_5$ HT + $Ni_{12}P_5$ LT + Ni_2P], which originates at the binary eutectoid e9, $Ni_{12}P_5$ HT = $Ni_{12}P_5$ LT + Ni_2P (994 °C), it forms the U-type reaction U13, $Ni_{12}P_5$ HT + Ni_2P = L + $Ni_{12}P_5$ LT, at 982 °C (see Fig. 5.9). This U-type reaction was only observed in samples in a quite limited concentration range, and the liquidus-apex of the reaction quadrangle was placed accordingly. Note that in this reaction the liquid phase appears on the product side on cooling, because there is only one three-phase field involving the liquid phase coming from higher temperature ([L + $Ni_{12}P_5$ HT + Ni_2P]), whereas two three-phase fields including the liquid phase are formed, i.e. [L + $Ni_{12}P_5$ LT + Ni_2P] and [L + $Ni_{12}P_5$ HT + $Ni_{12}P_5$ LT].

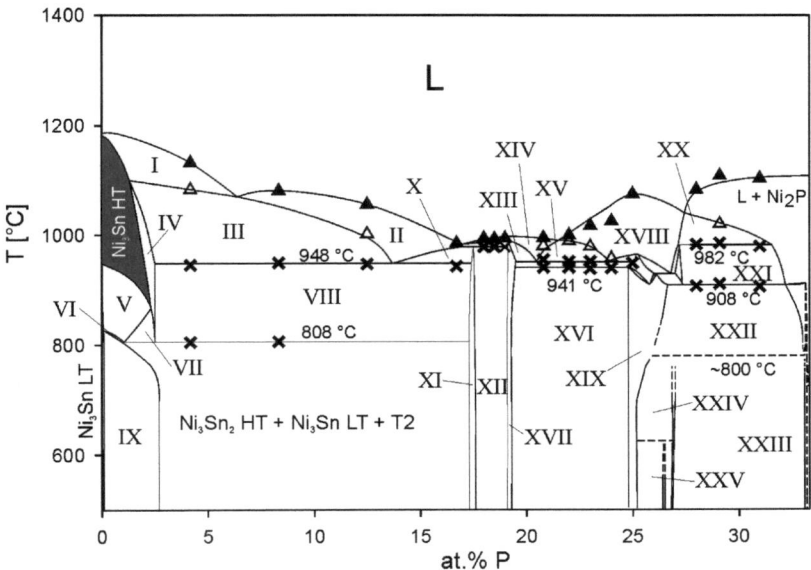

Fig.5.9: Isopleth from Ni_3Sn to Ni_2P with data points from thermal analysis. x: invariant effect, open triangle: monovariant effect, full triangle: liquidus. Phase field designations:

I	$L + Ni_3Sn$ HT	XIV	$L + T1$
II	$L + Ni_3Sn_2$ HT	XV	$L + Ni_{12}P_5$ LT + T1
III	$L + Ni_3Sn_2$ HT + Ni_3Sn HT	XVI	Ni_3Sn_2 HT + $Ni_{12}P_5$ LT + T1
IV	Ni_3Sn_2 HT + Ni_3Sn HT	XVII	Ni_3Sn_2 HT + T1
V	Ni_3Sn HT + Ni_3Sn LT	XVIII	$L + Ni_{12}P_5$ HT
VI	Ni_3Sn HT + Ni_3Sn LT + T2	XIX	Ni_3Sn_2 HT + $Ni_{12}P_5$ LT
VII	Ni_3Sn_2 HT + Ni_3Sn HT + Ni_3Sn LT	XX	$L + Ni_{12}P_5$ HT + Ni_2P
VIII	Ni_3Sn_2 HT + Ni_3Sn HT + T2	XXI	$L + Ni_{12}P_5$ LT + Ni_2P
IX	Ni_3Sn LT + T2	XXII	Ni_3Sn_2 HT + $Ni_{12}P_5$ LT + Ni_2P
X	$L + Ni_3Sn_2$ HT + T2	XXIII	$Ni_{12}P_5$ LT + Ni_2P + T3
XI	Ni_3Sn_2 HT + T2	XXIV	Ni_3Sn_2 HT + $Ni_{12}P_5$ LT + T3
XII	Ni_3Sn_2 HT + T1 + T2	XXV	Ni_3Sn_2 HT + $Ni_{12}P_5$ LT + T4
XIII	$L + Ni_3Sn_2$ HT + T1		

The three-phase field [$L + Ni_{12}P_5$ HT + $Ni_{12}P_5$ LT] runs into another U-type reaction, U6, $L + Ni_{12}P_5$ HT = Ni_3Sn_2 HT + $Ni_{12}P_5$ LT, which was only observed in the DTA recording of sample NPS 55 overlapping with other effects, because its reaction plane covers only a few at.% in P-concentration and no other sample was placed there. Based on this measurement it was placed in a consistent way at approx. 930 °C.

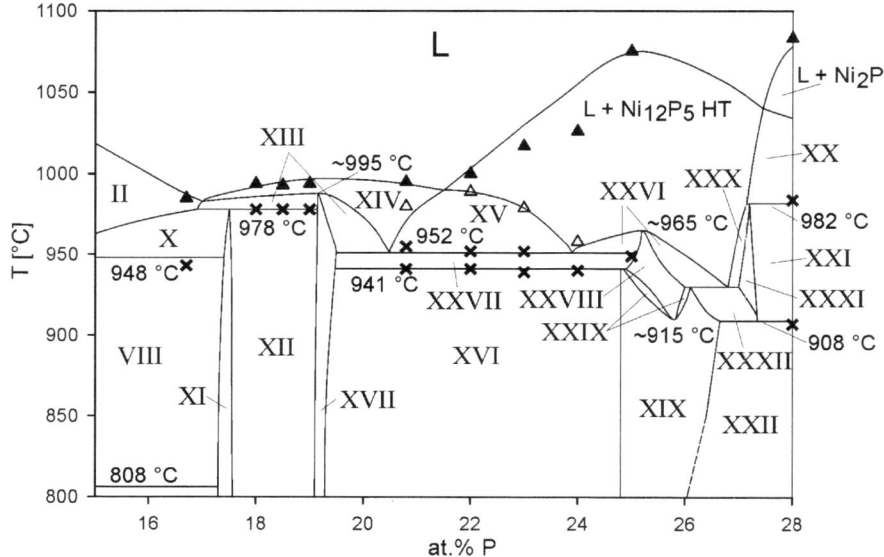

Fig. 5.10: Detail of the isopleth from Ni_3Sn to Ni_2P between 15 and 28 at.% P. The development of the phase equilibria around the close lying reactions U12, E4 and U7 as well as e15 and e16 related to the $Ni_{12}P_5$ HT – LT transition are shown in this Figure. Symbols as in Fig. 5.11. Phase field designations:

II	$L + Ni_3Sn_2$ HT	XIX	Ni_3Sn_2 HT + $Ni_{12}P_5$ LT
VIII	Ni_3Sn_2 HT + Ni_3Sn HT + T2	XX	$L + Ni_{12}P_5$ HT + Ni_2P
X	$L + Ni_3Sn_2$ HT + T2	XXI	$L + Ni_{12}P_5$ LT + Ni_2P
XI	Ni_3Sn_2 HT + T2	XXII	Ni_3Sn_2 HT + $Ni_{12}P_5$ LT + Ni_2P
XII	Ni_3Sn_2 HT + $Ni_{10}P_3Sn$ + T2	XXVI	$L + Ni_3Sn_2$ HT + $Ni_{12}P_5$ HT
XIII	$L + Ni_3Sn_2$ HT + T1	XXVII	Ni_3Sn_2 HT + $Ni_{12}P_5$ HT + T1
XIV	$L + T1$	XXVIII	Ni_3Sn_2 HT + $Ni_{12}P_5$ HT
XV	$L + Ni_{12}P_5$ LT + T1	XXIX	Ni_3Sn_2 HT + $Ni_{12}P_5$ HT + $Ni_{12}P_5$ LT
XVI	Ni_3Sn_2 HT + $Ni_{12}P_5$ LT + T1	XXX	$L + Ni_{12}P_5$ HT + $Ni_{12}P_5$ LT
XVII	Ni_3Sn_2 HT + T1	XXXI	$L + Ni_{12}P_5$ LT
		XXXII	$L + Ni_3Sn_2$ HT + $Ni_{12}P_5$ LT

The reaction sequence described so far (compare also the Scheil Diagram in Fig. 5.5) creates the three three-phase fields [$L + Ni_3Sn_2$ HT + Ni_2P] (originating at the quasi-binary eutectic e14), [$L + Ni_3Sn_2$ HT + $Ni_{12}P_5$ LT] (from e16 and U6) and [$L + Ni_{12}P_5$ LT + Ni_2P] (product of U13). They merge in the ternary eutectic E3, L = Ni_3Sn_2 HT + $Ni_{12}P_5$ LT + Ni_2P at 908 °C (see Fig. 5.6). The extremely fine matrix seen in the micrograph of sample NPS 56 in Fig. 5.13 (page 89) was interpreted to be the product of this ternary eutectic reaction. The micrograph also shows the primary crystallization of Ni_2P and the secondary crystallization of $Ni_{12}P_5$ (HT in this case, which transformed into LT during cooling). This ternary eutectic reaction represents the lowest point of the liquidus surface in the triangle Ni$_3$Sn$_2$ HT – Ni$_{12}$P$_5$ – Ni$_2$P (local minimum, see Figs. 5.9, 5.11 and 5.12).

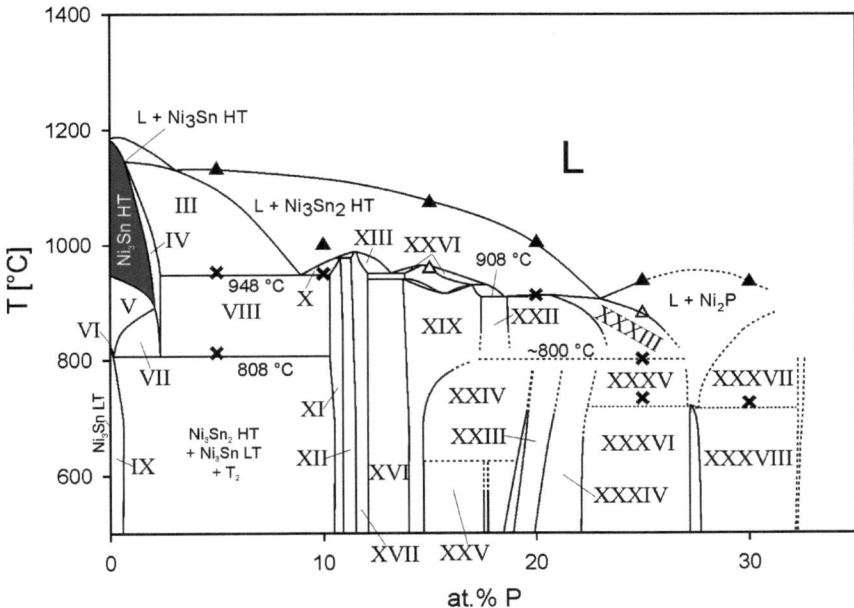

Fig. 5.11: Partial Isopleth from Ni$_3$Sn to P until 40 at.% P. Symbols as in Fig. 5.1. Phase field designations:

III	L + Ni$_3$Sn$_2$ HT + Ni$_3$Sn HT	XIX	Ni$_3$Sn$_2$ HT + Ni$_{12}$P$_5$ LT
IV	Ni$_3$Sn$_2$ HT + Ni$_3$Sn HT	XXII	Ni$_3$Sn$_2$ HT + Ni$_{12}$P$_5$ LT + Ni$_2$P
V	Ni$_3$Sn HT + Ni$_3$Sn LT	XXIII	Ni$_{12}$P$_5$ LT + Ni$_2$P + T3
VI	Ni$_3$Sn HT + Ni$_3$Sn LT + T2	XXIV	Ni$_3$Sn$_2$ HT + Ni$_{12}$P$_5$ LT + T3
VII	Ni$_3$Sn$_2$ HT + Ni$_3$Sn HT + Ni$_3$Sn LT	XXV	Ni$_3$Sn$_2$ HT + Ni$_{12}$P$_5$ LT + T4
VIII	Ni$_3$Sn$_2$ HT + Ni$_3$Sn HT + T2	XXVI	L + Ni$_3$Sn$_2$ HT + Ni$_{12}$P$_5$ HT
IX	Ni$_3$Sn LT + T2	XXXIII	L + Ni$_3$Sn$_2$ HT + Ni$_2$P
X	L + Ni$_3$Sn$_2$ HT + T2	XXXIV	Ni$_2$P + T3
XI	Ni$_3$Sn$_2$ HT + T2	XXXV	L + Ni$_2$P + T3
XII	Ni$_3$Sn$_2$ HT + T1 + T2	XXXVI	Ni$_2$P + T3 + T5
XIII	L + Ni$_3$Sn$_2$ HT + T1	XXXVII	L + Ni$_5$P$_4$ + Ni$_2$P
XVI	Ni$_3$Sn$_2$ HT + Ni$_{12}$P$_5$ LT + T1	XXXVIII	Ni$_2$P + Ni$_5$P$_4$ + T5
XVII	Ni$_3$Sn$_2$ HT + T1		

On the Ni-rich side of reaction e16 (L = Ni$_3$Sn$_2$ HT + Ni$_{12}$P$_5$ HT) the respective three-phase field opens up again and ends in a eutectic reaction, E4: L = Ni$_3$Sn$_2$ HT + Ni$_{12}$P$_5$ HT + T1 at 952 °C, which was clearly observed in thermal analysis (Figs. 5.6, 5.9 and 5.10). Additional weak thermal effects with an onset at 941 °C were noticed in the DTA recordings of the same samples in this area (see Figs. 5.9 and 5.10). Due to their considerably lower intensity compared to E4, these effects were interpreted as the solid state reaction U7: $Ni_{12}P_5$ HT + T1 = Ni_3Sn_2 HT + $Ni_{12}P_5$ LT, which occurs only 11 °C below eutectic E4. From there the three-phase field [Ni_3Sn_2 HT + $Ni_{12}P_5$ HT + $Ni_{12}P_5$ LT] continues to the quasi binary eutectoid reaction e15 (~915°C), where it merges with its other branch coming from U6. This sequence thus completes the transition from $Ni_{12}P_5$ HT to $Ni_{12}P_5$ LT in the ternary and is shown in the vertical sections in Figs. 5.9-5.12.

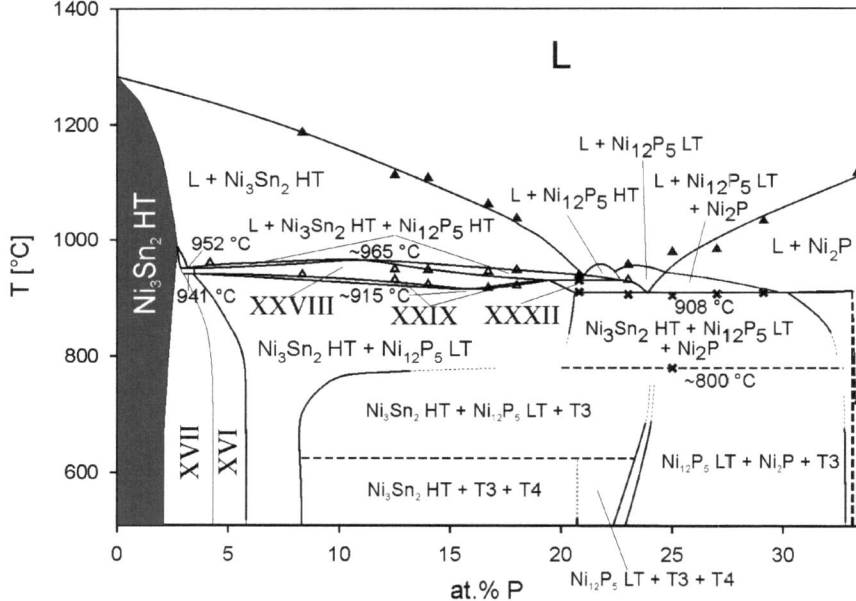

Fig. 5.12: Isopleth from Ni_3Sn_2 HT to Ni_2P. Symbols as in Fig. 5.11. Phase field designations:

XVI	Ni_3Sn_2 HT + $Ni_{12}P_5$ LT + T1	XXIX	Ni_3Sn_2 HT + $Ni_{12}P_5$ HT + $Ni_{12}P_5$ LT
XVII	Ni_3Sn_2 HT + T1	XXXII	L + Ni_3Sn_2 HT + $Ni_{12}P_5$ LT
XXVIII	Ni_3Sn_2 HT + $Ni_{12}P_5$ HT		

Of course, in reaction E4 (952 °C) two further three-phase fields are involved, i.e. [L + T1 + $Ni_{12}P_5$ HT] and [L + $Ni_{10}P_3Sn$ + Ni_3Sn_2 HT]. Both these phase fields emerge in quasi-binary reactions, too, that involve the ternary compound T1. This phase is the ternary compound with the highest congruent melting point (1010 °C) in the Ni-P-Sn system (cf. Chapter 5.1) and participates in a total of three quasi-binary eutectic reactions: e19: L = T1 + Ni_3Sn_2 HT, e18: L = T1 + $Ni_{12}P_5$ HT and e17: L = T1 + $Ni_{21}P_6Sn_2$. As the temperatures of these reactions were not observed directly in thermal analysis, they were placed at temperatures that are consistent with the surrounding phase equilibria, i.e. 988, 995 - 1010 and 978 - 991 °C. However, these reactions play an important role as starting points for the development of the phase equilibria in this part of the phase diagram. There are basically two branches of liquidus valleys together with their related sequences of invariant reactions, one originating from e18, L = T1 + $Ni_{12}P_5$ HT, the other one from e19, L = T1 + Ni_3Sn_2 HT (compare the reaction scheme in Fig. 5.5 and the liquidus projection in Fig. 5.6).

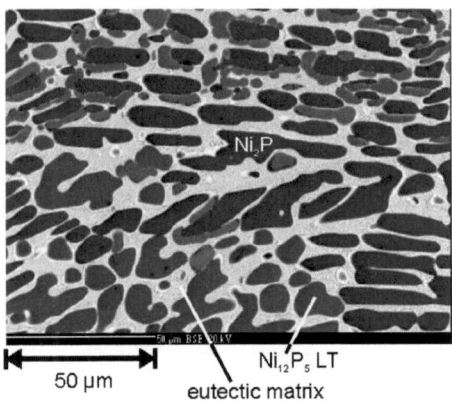

Fig. 5.13: SEM image of as cast sample NPS 56. The black grains of Ni_2P are considered the primary crystallization, wheras the dark grey grains are the secondary crystallization of $Ni_{12}P_5$ HT (LT in XRD). The fine matrix was interpreted to contain the three phases Ni_3Sn_2 HT + $Ni_{12}P_5$ LT + Ni_2P.

The following reaction sequence starting from the quasi binary eutectic e18 (995 - 1010 °C) and from binary reactions is suggested here for the vicinity of the Ni_5P_2 and $Ni_{12}P_5$ phases. It can be best described as a sequence of invariant U-type reactions. The highest ternary reaction is U16: L + Ni_5P_2 HT = Ni_5P_2 LT + $Ni_{12}P_5$ HT, which is connected to the binary reactions p9: L + $Ni_{12}P_5$ HT = Ni_5P_2 HT and p8 : L + Ni_5P_2 HT = Ni_5P_2 LT. The three-phase fields [L + T1 + $Ni_{12}P_5$ HT] and [L + T1 + T2] originating at the respective quasi-binary reactions e18 and e17, both join into U-type reactions: U15, L + $Ni_{12}P_5$ HT = T1 + Ni_5P_2 LT and U14 : L + T1 = Ni_5P_2 LT + T2, respectively. The temperature of U15 was not observed by thermal analysis, while U14 was placed at 985 °C according to the thermal effects in the DTA curves of samples in the proper concentration range. The further continuation of the reaction sequence, i.e. any further reaction involving the three-phase field [L + Ni_5P_2 LT + T2] (product of U14) could not be clarified in this work. This is due to the fact that in sample NPS 79b, $Ni_{73}P_{25}Sn_2$, EPMA measurements revealed a phase having the composition $Ni_{72.9}P_{26.3}Sn_{0.8}$ (labelled "T", see Chapter 5.2). Thermal effects that may indicate an invariant reaction related to the formation of another ternary compound were indeed observed, but they are too inconclusive to allow for a definite interpretation. Therefore, it is currently not clear whether this composition represents another ternary compound or a solid solution of Ni_5P_2 LT or even HT and this part has been left open in Figs. 5.5 and 5.6.

The reaction sequence described here (neglecting the omitted part) continues via reaction U8: L + T = Ni_3P + T2 at 947 °C and terminates in a eutectic reaction at 861 °C: E2: L = (Ni) + Ni_3P + T2.

The other three-phase field resulting from U16 (see above), Ni_5P_2 HT + Ni_5P_2 LT + $Ni_{12}P_5$ HT, merges with another one coming from the binary reaction e10 at approx. 1005 °C (see Fig. 5.5) in a solid state reaction U11: Ni_5P_2 HT + $Ni_{12}P_5$ HT = Ni_5P_2 LT + $Ni_{12}P_5$ LT (between 987 and 1005 °C). From there [Ni_5P_2 LT + $Ni_{12}P_5$ HT + $Ni_{12}P_5$ LT] continues to U10: Ni_5P_2 LT + $Ni_{12}P_5$ HT = T1 + $Ni_{12}P_5$ LT. This latter reaction was placed at 960 °C according to the thermal effects observed in the DTA, and is connected to U7 at 941 °C via the three-phase field [$Ni_{12}P_5$ HT + $Ni_{12}P_5$ LT + T1].

The other branch starts on the Ni-rich side of reaction e19, where the corresponding three-phase field [L + Ni_3Sn_2 HT + T1] merges with that from L = T1 + T2 (e17) to form reaction U12: L + T1 = Ni_3Sn_2 HT + T2 at 978 °C (Fig. 5.10). From there yet again a sequence of reactions continues via U9: L + Ni_3Sn_2 HT = Ni_3Sn HT + T2 at 948 °C and U5 : L + Ni_3Sn HT = (Ni) + T2 at 872 °C to the eutectic reaction E2. U9 and U5 are linked to eutectic reactions in the binary Ni-Sn system, e12 and e13, while in the ternary eutectic reaction E2 the two described reaction branches merge. E2 also represents the lowest liquidus temperature in the Ni-rich corner of the Ni-P-Sn system (see Figs. 5.5 and 5.6).

All phase equilibria discussed so far are on the Ni-rich side of the quasi binary reaction e14. On the (P,Sn)-rich side of this reaction the three-phase field [L + Ni_2P + Ni_3Sn_2 HT] opens again, and was as such observed in the isotherm at 850 °C (Fig. 5.1). This three-phase field gains further importance due to the fact that it forms the continuation of the phase equilibria to the Sn- and P-rich parts of the phase diagram. A full evaluation of these parts of the system is currently in progress [95].

Thermal effects observed at approx. 800 °C in some samples placed between Ni_3Sn_2 HT and Ni_2P (e.g. NPS 29, see also Table 5.3) were interpreted to be related to the formation of the T3 phase ($Ni_{10}P_3Sn_5$) (see also Chapter 5.3). This temperature nicely agrees with 797 °C mentioned by Furuseth and Fjellvåg [36] as the lower stability limit of the Ni_3Sn_2 HT ternary solid solution at a composition of $Ni_{53.5}P_{16.3}Sn_{30.2}$ (close to the Sn-rich stability limit determined for T3 by EPMA).

The exact formation mechanism of T3 out of Ni_3Sn_2 HT has so far not been elucidated. Ni_3Sn_2 HT crystallizes in a partially filled $InNi_2$ type structure ($B8_2$-type, derived from the NiAs type). These B8 type crystal structures are extremely flexible with respect to ordering of atoms and/or vacancies and the filling of interstitial positions with additional atoms. Thus, huge homogeneity ranges and the formation of super structures are frequent, e.g. in Ni_3Sn_2 and Cu_6Sn_5.

In the Ni-P-Sn system the formation of the phases T3 and T4 out of the solid solution of Ni_3Sn_2 HT has to be a first order transition, because there is no direct group / subgroup relation between the space groups $P6_3/mmc$ and $P\overline{1}$. However, due to the crystallographic flexibility mentioned above, these transitions are often quite complex and can involve a succession of phases with gradually lower symmetry. Because of these crystallographic complexities that are most likely reflected by the phase diagram the formation of the T3 phase has not been investigated in detail in this work. However, Ni_2P can be expected to be involved in the reaction at app. 800 °C, because the relevant thermal effects were found in samples between Ni_3Sn_2 HT and Ni_2P. In the vertical sections (Figs. 5.9, 5.11 and 5.12) a dashed line is shown at 800 °C, but the type of transition is not indicated.

With respect to the formation of T3 and T4, a number of solid state reactions have to occur between 700 and 800 °C in order to form the phase triangulation between [Ni_2P + T3] and [Ni_3Sn_2 HT + $Ni_{12}P_5$ LT + T3] at 700 °C (cf. the phase diagram sections in this study, Figs. 5.1-5.3, 5.5, 5.6 and 5.9-5.12). However, due to the fact that no thermal effects related to these reactions have so far been observed by DTA, the temperature range between 850 and 700 °C is left open or shown by dotted lines in the corresponding composition range.

A further solid state reaction has to exist between 550 and 700 °C, U1, Ni_3Sn_2 HT + T3 = $Ni_{12}P_5$ LT + T4, in order to yield the three-phase fields [Ni_3Sn_2 HT + $Ni_{12}P_5$ LT + T4] and [$Ni_{12}P_5$ LT + T3 + T4] that were observed at 550 °C.

In Chapter 5.2 the remarkable appearance of the Ni_3Sn HT-phase at 850 °C was discussed and the temperature range for the transition to the LT-phase was given to be 808 to 830 °C in the ternary system. In agreement with the other phase equilibria this transition comprises a U type reaction, U4: (Ni) + Ni_3Sn HT = Ni_3Sn LT + T2 and a eutectoid decomposition of the HT phase, E1: Ni_3Sn HT = Ni_3Sn LT + Ni_3Sn_2 HT + T2. The temperature of the eutectoid reaction was determined to be 808 °C, while the temperature of U4 is based on a thermal effect at approx. 830 °C observed in sample NPS 64. As both reactions only involve solid phases, their signals in the DTA curves are weak.

In the vicinity of composition "T" a solid state Type II reaction: U3: T + T2 = Ni_3P + T1 was proposed in Chapter 5.3. However, no thermal effects pertinent to this reaction were found in the

DTA measurements. Thus the corresponding change in the phase equilibria between 850 and 700 °C is the only basis for the inclusion of this reaction.

Table 5.6: Preliminary results of the phase analysis in the system P-Sn

No.	Nominal Composition [at %]	Heat Treatment [°C]	Phase Analysis			WDS [at.%]		Σ wt.%
			Phase	Structure Type	Lattice Param. [pm]	P	Sn	
PS 2	$P_{10}Sn_{90}$	500, 38d	(Sn)	βSn	no XRD made	0.0	100.0	99.4
			P_3Sn_4	Bi_3Se_4		41.2	58.8	98.7
PS 4	$P_{35}Sn_{65}$	500, 38d	(Sn)	βSn	a=583.11(4) c=318.08(3)	0.0	100.0	98.8
			P_3Sn_4	Bi_3Se_4	a=396.788(7) c=3533.49(9)	41.5	58.5	99.4
PS 9	$P_{40}Sn_{60}$	500, 32d	(Sn)	βSn	a=583.25(7) c=318.07(5)	41.8	58.2	100.5
			P_3Sn_4	Bi_3Se_4	a=396.92(1) c=3534.3(2)		not found in EPMA	
PS 5	$P_{44.4}Sn_{55.6}$	500, 38d	P_3Sn_4	Bi_3Se_4	a=396.743(5) c=3533.60(6)	41.2	58.8	99.4
			P_4Sn_3	P_4Sn_3	a=443.07(2) c=2838.5(2)		not found in EPMA	
PS 6	$P_{50}Sn_{50}$	500, 21d	P_3Sn_4	Bi_3Se_4	a=396.818(5) c=3534.19(6)		not determined	
			P_4Sn_3	P_4Sn_3	a=443.16(5) c=2837.4(6)			
PS 10	$P_{55}Sn_{45}$	500, 32d	P_3Sn_4	Bi_3Se_4	a=396.90(4) c=3533.6(7)	42.1	57.9	99.7
			P_4Sn_3	P_4Sn_3	a=443.131(6) c=2839.20(6)	56.3	43.7	100.4
PS 14	$P_{65}Sn_{35}$	500, 27d	P_3Sn_4	Bi_3Se_4	a=396.84(1) c=3533.6(4)		not determined	
			P_4Sn_3	P_4Sn_3	a=443.108(7) c=2839.14(7)			
			P_3Sn		a=736.59(2) c=1052.05(6)			
PS 15	$P_{70}Sn_{30}$	500, 27d	P_3Sn_4	Bi_3Se_4	a=396.96(3) c=3533.6(5)		not determined	
			P_4Sn_3	P_4Sn_3	a=443.16(2) c=2839.5(2)			
			P_3Sn		a=736.69(2) c=1052.26(4)			

Table 5.7: Experimental results of the thermal analysis in the system P-Sn.

No.	Nominal Comp. [at.%]	Heat Treatm. [°C]	Conditions	Thermal Analysis Heating [°C] Invariant Effects	Heating [°C] Other Effects	highest effect on heating	Cooling [°C] highest effect on cooling
PS 1	P_2Sn_{98}	200, 2d	Sn-Reference, 2 K/min,	226 (from sample), 230 (from reference) Note: no temperature calibration		---	---
				this measurement was carried out to identify the type of the most Sn-rich reaction; see main text			
PS 2	$P_{10}Sn_{90}$	500, 38d	5 K/min,	232		503	458
PS 3	$P_{20}Sn_{80}$	500, 38d	5 K/min,	232		549	505
PS 4	$P_{35}Sn_{65}$	500, 38d	5 K/min,	231	558 eo	570	498
PS 9	$P_{40}Sn_{60}$	500, 32d	5 K/min,	230	569 eo	580	514
PS 5	$P_{44.4}Sn_{55.6}$	500, 38d	5 K/min,		566 eo, 575 eo	579	496
PS 6	$P_{50}Sn_{50}$	500, 21d	5 K/min,			575	524
PS 10	$P_{55}Sn_{45}$	500, 32d	5 K/min,		554 eo, 571 eo	575	519

eo = extrapolated onset

5.5 A brief note on the P-Sn system

In the ternary Ni-P-Sn system phase equilibria in the Sn-rich part were investigated at 550 and 200 °C, for which knowledge of the binary P-Sn phase diagram is crucial. Although an apparently complete phase diagram description is available from Refs. [76] and [44], the poor amount of further phase diagram related literature and the lack of EPMA and DTA studies make a new investigation in this system worthwhile (see Chapter 2.3). Due to experimental difficulties caused by heavy evaporation and loss of P during equilibrium annealing, work on this system is still very much in progress. However, preliminary data for the Sn-rich part can be given: phase analysis in Table 5.6 and thermal data in Table 5.7.

Three binary compounds as described in the literature were found in the present work: P_3Sn_4, P_4Sn_3 and P_3Sn. The compositions of P_3Sn_4 ($P_{41.5}Sn_{58.5}$) and P_4Sn_3 ($P_{56.3}Sn_{43.7}$) as determined by EPMA agree with the respective literature data. While P_3Sn_4 was found to be a line compound, only the Sn-rich phase boundary of P_4Sn_3 could so far be determined. In samples PS 14 and 15 on the P-rich side of P_4Sn_3 all three compounds were found, which points to non-equilibrium, and therefore no reliable EPMA data could be produced so far.

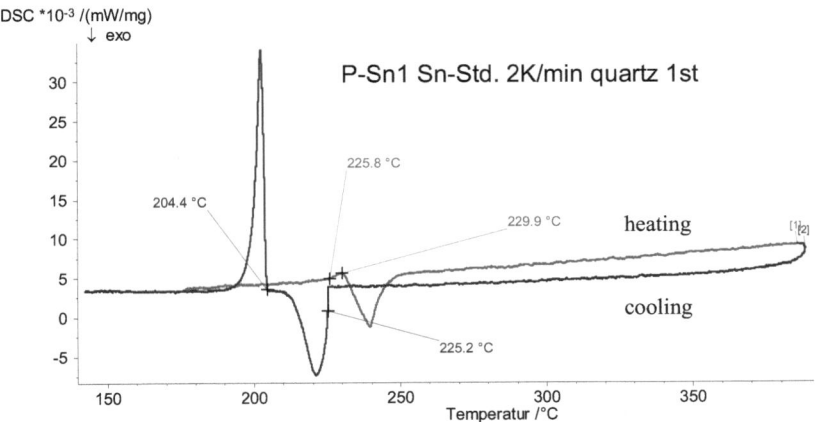

Fig. 5.14: DTA recording of sample P_2Sn_{98} carried out at 2K/min with Sn as reference. The sequence of peaks in the heating curve (green) indicates a eutectic reaction.

DTA data obtained in the Sn-rich part of the P-Sn system at P-concentrations < 23 at.% revealed the existence of an invariant reaction at 231 °C. According to Vivian [76] this reaction takes place at the melting temperature of pure Sn (232 °C), but the author did not specify the type of this reaction.

In order to distinguish between a eutectic and a peritectic, a sample P_2Sn_{98} (PS 1) was measured by DTA using pure Sn as a reference. The melting of the Sn reference results in a peak pointing in the "wrong", i.e. exothermic direction. The sequence of reference and sample effects thus allows a determination of the reaction type. In case of a eutectic reaction the proper endothermic sample effect should occur before the reference peak pointing in the "exothermic" direction.

Fig. 5.14 shows the DTA recording at 2K/min of sample PS 1 with Sn as reference. In the heating curve the effects of sample and reference overlap. It is, however, clearly visible that the peak onset points into the endothermic direction, while the rest of the effect points into the exothermic direction. This sequence, of course, indicates a eutectic reaction.

Based on this result the most Sn-rich reaction was assigned the eutectic L = (Sn) + P_3Sn_4. The temperature of this reaction was determined to be 231 °C from other samples, because no temperature calibration could be used for the measurement described above.

The liquidus in the region of P < 23 at.% (at compostions more Sn-rich than the L1 + L2 miscibility gap) agrees with the literature phase diagram.

The further DTA measurements, however, indicate a couple of ambiguities in this system. The thermal effects listed in Table 5.7 do not show effects at the same temperature in various DTA recordings that could correspond to the syntectic L1 + L2 = P_3Sn_4 (550 °C according to [76]). Samples placed in the L1 + L2 region and annealed at 650 °C for metallographic investigation did not yield useful results due to the loss of P and the resulting composition shift (more than 10 at.%). It appears that samples heated above the liquidus lose tremendous amounts of P by evaporation.

With respect to the ternary system it can be resumed that relevant data from the binary P-Sn system are still missing or unclear. Results in the Sn-rich corner of the ternary Ni-P-Sn system therefore have tentative character, as long as a full study of the binary P-Sn system has not been carried out.

5.6 Sn-rich phase equilibria

The full isothermal section at 550 °C including the Sn-rich part is shown in Fig. 5.3a (Chapter 5.3, page 69). Phase equilibria were deduced from the phase analyses of samples listed in Table 5.1. While most of the phase fields are based on the analyses of several Sn-rich samples, the most P-rich phase fields (shown by dashed lines) have been added in a consistent way, but do not have any experimental foundation. Also note that samples NPS 93-99 were intended to have a P content of 60 at.%, but shifted during annealing due to P-loss towards the (Ni,Sn)-side. Particularly high weight loss was observed for the samples placed closer to the (P,Sn)-rich side, which is consistent with the observation of higher weight losses in binary P-Sn samples. The new nominal compositions of these samples were calculated from the assumption that only P had been lost during heat treatment.

The liquid phase field shown in Fig. 5.3a was estimated from thermal effects observed in DTA recordings. In the binary P-Sn system the syntectic reaction L1 + L2 = P_3Sn_4 has been proposed in Ref. [76] at 550 °C. As the extent of the related miscibility gap both in the binary and the ternary systems is currently not established, it has not been considered for the present course of the liquidus.

From the XRD and EPMA data can be seen that the P_3Sn_4 phase appears in most of the samples quenched from 550 °C either as an equilibrium phase or as a product of liquid decomposition during quenching. These two cases could be distinguished from the micrographs of the respective samples. For example, Fig. 5.15 shows the micrographs of samples NPS 72 and 77 quenched from 550 °C.

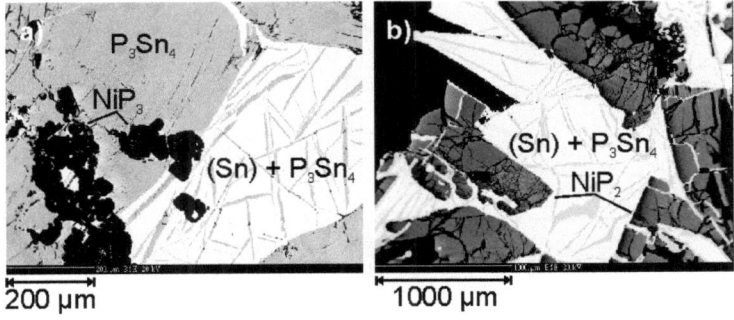

Fig. 5.15: SEM images of samples a) NPS 72 and b) NPS 77 quenched after annealing at 550 °C. In a) large grains of P_3Sn_4 and smaller ones of NiP_3 as well as the matrix from (Sn) and P_3Sn_4 can be seen. b) shows NiP_2 and the matrix from (Sn) and P_3Sn_4 (the black areas are holes). While the large grains indicate that P_3Sn_4 is an equilibrium phase at the annealing temperature, P_3Sn_4 in the matrix is a product of decomposition of the liquid during quenching.

The large grains of P_3Sn_4 in the micrograph of sample NPS 72 indicate that this phase is an equilibrium phase at the annealing temperature, while in sample NPS 77 P_3Sn_4 was only found in the matrix as a product of liquid decomposition. As this decomposition of the liquid prevents the determination of the L-apexes of the three-phase field from EPMA data, this information helped to establish the phase triangulation in Fig. 5.3a.

EPMA data obtained from ternary Ni-P-Sn samples annealed at 550 °C for the binary P_3Sn_4 phase, which is a line compound in the binary P-Sn system show that there is no significant solubility of Ni in this phase, too.

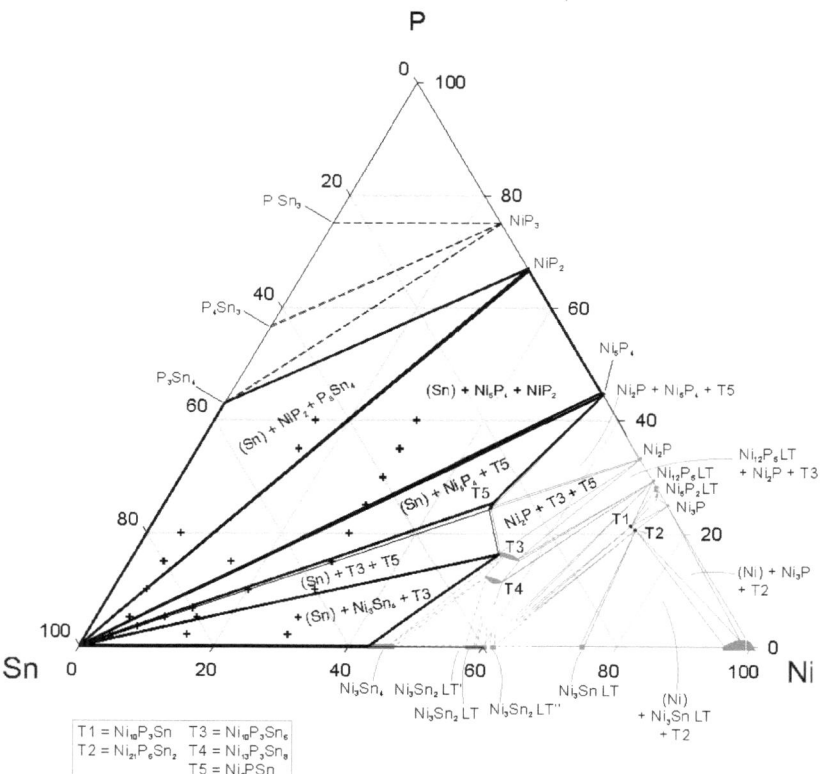

Fig. 5.16: Partial isothermal section at 200 °C. The part that was adapted from the phase equilibria at 550 °C is shown in grey.

At 200 °C only the Sn-rich part is a-priori experimentally accessible in a system of otherwise high-melting or volatile elements because of slow equilibration in the high melting areas. Only samples that were more Sn-rich than Ni_3Sn_4 were annealed at this temperature, and the remaining phase equilibria have been adapted from 550 °C (they are therefore shown in grey). The partial isothermal section at 200 °C derived from these samples is shown in Fig. 5.16.

In the binary Ni-Sn and P-Sn systems the solubility of Ni or P in Sn is insignificant at 200 °C. The same situation was found for (Sn) in the ternary system, too, where EPMA / EDX did not reveal any appreciable homogeneity range. The region of the Ni_3Sn_2 LT, LT' and LT'' was not investigated and has therefore been left open in the isotherm at 200 °C (Fig. 5.16).

Otherweise, samples annealed at 200 °C were prone to by a high number of experimental problems. Frequently, more than the allowed three phases were found. This can be explained by several reasons:

- the appearance of Ni_2P (parent phase of T5 - Ni_2PSn) due to incomplete formation of T5
- extension of metastable reactions from the binary Ni-P system into the ternary
- influence of the gas phase
- low annealing temperature

The combination of all of these effects made interpretation of the primary results complicated, which is nicely demonstrated by the observation of five phases, e.g. in samples NPS 66 and 76, or four phases, e.g. in samples NPS 75 and NPS 84 - 86. Fig. 5.17 shows part of the microstructures of samples NPS 66 and 75. In the image of sample NPS 66 grains of Ni_2P completely surrounded by T5 can be seen, which indicates two things:

- the transformation to T5 is incomplete
- T5 is formed out of Ni_2P and the liquid in a (quasi-binary) peritectic reaction

Pre-annealing the samples at 700 °C just below the reaction temperature of 722 °C of L + Ni_2P = T5 for two weeks reduced the amount of Ni_2P that was finally found after equilibrium annealing, but Ni_2P could not be removed entirely.

However, the microstructure reveals the solidification behaviour of the samples, and the peritectic rim of T5 around Ni_2P can clearly be seen. Thus the microstructure supports the existence of a quasi-binary peritectic formation of T5.

This feature can also be found in the microstructure of sample NPS 75 (Fig. 5.17b). This sample, however, shows a total of four different phases and is therefore useless for the determination of the phase triangualtion. The appearance of such a high number of phases in one single sample was found to be common in samples placed in the phase fields [(Sn) + Ni_5P_4 + T5] and [(Sn) + Ni_5P_4 + NiP_2]. These phase fields were therefore placed in consistency with the data from surrounding samples and data from other temperatures.

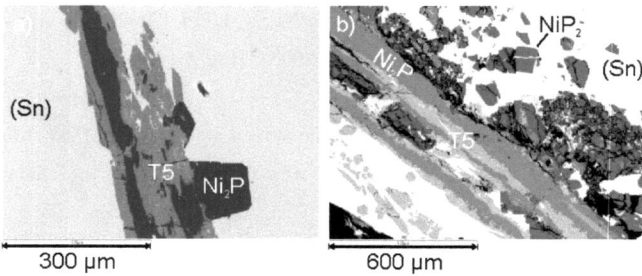

Fig. 5.17: SEM images of samples a) NPS 66 and b) NPS 75 quenched from 200 °C. Non-equilibrium effects due to incomplete reaction L + Ni$_2$P = T5 can be seen in both micrographs. Furthermore sample 75 contains four phases that do not correspond to the phase triangulation, either.

Furthermore, the four phases found in sample NPS 75 do not correspond to the phase triangulation. Generally, interpretation of the experimental results at 200 °C was further aggravated by the fact that results that were apparently derived from equilibrium samples violate the clear cross-principle, i.e. contradicting phase fields could be derived from different samples; e.g. compare the contradicting results in samples NPS 75 and NPS 86-89 in Table 5.1. From these samples the phase fields [(Sn) + Ni$_5$P$_4$ + T5] or [(Sn) + NiP$_2$ + T5] would be possible. A similar situation was encountered for the phase fields [(Sn) + Ni$_3$Sn$_4$ + T3] or [(Sn) + Ni$_3$Sn$_4$ + T5], respectively.

This ambiguous situation was sorted out using DTA data obtained from the respective samples, which below 550 °C only show thermal effects at ~230-232 °C. These effects indicate the existence of ternary reactions that are likely to be linked to the binary Sn-rich eutectics in Ni-Sn and P-Sn. The lack of other thermal effects suggests that the principal phase triangulation should remain the same when going from 550 to 200 °C (except for the solidification of Sn, of course). The phase triangulation shown in the isothermal section at 200 °C (Fig. 5.16) has been based on this fact.

5.7 Conclusion and Lessons Learned from the Ternary Phase Diagram

Out of the phase equilibria described between 200 °C and the liquidus in the Ni-rich corner, only those at 200 °C seem to have direct relevance for soldering. In Chapters 1.5 and 2.4 two unclear issues from the literature were mentioned, i.e. the appearance of a solid solution of P in Ni$_3$Sn$_2$ in solder joints and the existence of a ternary compound Ni$_3$PSn. In the present study answers can be given to these issues: the solid solution of P in Ni$_3$Sn$_2$ was only found at 850 °C, while at the lower temperatures only a minor solubility was found, which is unlikely to increase when going from 550 to 200 °C (cf. Figs. 5.3a,b and 5.16).

The compound Ni_3PSn was not found in this study. According to the present experiments it does not exist in the temperature range from 550 to 850 °C. It may, however, exist in the Ni-rich part of the phase diagram at 200 °C, where no samples were placed because equilibrium is likely not to be reached due to the low annealing temperatures compared to the liquidus temperature.

This, however, means that in the most interesting region of the phase diagram at the temperature most relevant for soldering no reliable experimental information can be provided. Similarly, it has to be concluded, that the current version of the Sn-rich phase equilibria is still rather tentatitive due to the high amount of experimental difficulties (e.g. high P vapour pressure, P-loss, non-equilibrium, etc.).

Additional problems concerned the Ni_3Sn_2 HT phase at 700 °C, where SEM revealed four phases in samples 100 – 102. On cooling (quenching) the phases present at the annealing temperature seem to have further decomposed. This, however, is a serious problem, because it puts isothermal sections obtained from annealed and quenched samples under question. Precipitations indicate an unsuccessful quenching procedure, so that even the measured compositions should be considered with caution.

All these difficulties show that it is not sufficient to simply characterize a complex intermetallic system at one temperature alone, even if the prime technological interest may be focussed at this temperature (e.g. 200 or 250 °C for (lead-free) soldering). Unclear results and inconsistent data can only be recognized and sorted out by the use of many different methods and information from various temperatures and parts of the phase diagram.

In complex systems the determination of the phase equilibria by experimental techniques alone will probably not lead to satisfying results. A multidisciplinary approach involving e.g. the CALPHAD type calculations, too, is thus required, especially if volatile elements like P are included, where the gas phase has to be taken into account for the phase equilibria. Furthermore, experimental techniques beyond the standard set of XRD, EPMA and DTA should be employed to produce additional information, e.g. the diffusion couple technique. The phase diagram version for the Ni-P-Sn system proposed in this thesis should therefore be used as starting point for further experimental work, for which many points have been highlighted, and for modelling, which also will show regions, where further experimental work is necessary.

6. The Crystal Structure of C_6Cr_{23}-type $Ni_{21}Sn_2P_6$ (T2)

This Chapter is based on Ref. [100]. Copyright (2009); the full article including Figures and Tables has been reprinted with permission from Wiley-VCH.

6.1 An Overview over Selected Ternary Ordered C_6Cr_{23}-type Phases

The C_6Cr_{23} structure type is well known from literature, where it has been described in detail by numerous authors (cf. Refs. [101-103]). It is the basic structure type of the so called τ-carbides, which attracted attention due to their adverse effect on mechanical properties of Cr- and Fe-based alloys [103]. Ternary compounds of this type have also been reported in literature, e.g. $Fe_{21}W_2C_6$ by Westgren [104]. Moreover it was recognized that C can be replaced by other non-metals such as B or P (see e.g. Refs. [105-109]) resulting in a whole family of binary and ternary compounds. A detailed account of ternary C_6Cr_{23} type borides and their crystal chemistry is available from Stadelmayer et al. [107], who investigated a series of these compounds using XRD, among them $Ni_{21}Sn_2B_6$. According to these authors there are two ordered variants of ternary C_6Cr_{23} phases ($Fm\overline{3}m$):

Variant 1: $M'_{21}M''_2X_6$ with M'' only occupying position 8c
Variant 2: $M'_{20}M''_3X_6$ with M'' occupying positions 4a and 8c

The ternary phosphides $Ni_{20}Mg_3P_6$ and $Ni_{20}Mn_3P_6$ have been reported by Keimes and Mewis [108], while $Ni_{21}In_2P_6$ was described by Andersson-Soederberg and Andersson [109]. The existence of $Ni_{21}Sn_2P_6$ would be the logic extension of the above mentioned series. While earlier attempts to prepare this phase by Keimes and Mewis were not successful [108], work on the ternary Ni-P-Sn phase diagram (see Chapter 5) had indeed revealed the existence of this phase.

6.2 Description of the Crystal Structure and Discussion

Phase analysis based on Guinier-type measurements of several Ni-P-Sn alloys revealed the existence of the new phase $Ni_{21}Sn_2P_6$ at 550, 700 and 850 °C (see also Chapter 5.1). Due to the existence of both $Ni_{10}P_3Sn$ (T1) and $Ni_{21}Sn_2P_6$ (T2) at all investigated temperatures a high temperature – low temperature relation of the two phases can be ruled out. No significant variation of the lattice parameter of $Ni_{21}Sn_2P_6$ found in multi-phase samples with different nominal overall composition was found: from 1111.38 pm on the Ni-rich side to 1111.85 on the Sn-rich side at

550 °C and from 1111.40 to 1111.75 at 700 °C. This suggests an almost stoichiometric composition (line compound) – see also Chapter 5.1 and Tables 5.1 and 5.2.

The powder diffractogram, obtained from a single phase sample with the overall composition $Ni_{21}Sn_2P_6$, is given in Fig. 6.1. The observed diffraction pattern and the refinement using the crystal structure information from the single crystal analysis were found to be in good agreement (Rwp = 3.55 %). A minor difference was noticed for the lattice parameter, which was found to be 1112.2 pm for the single crystal refinement, and 1111.73 pm for the powder diffractometer refinement.

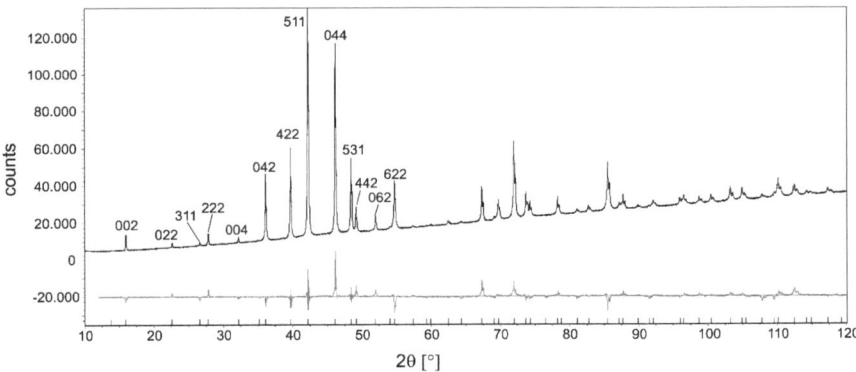

Fig. 6.1: Powder diffraction pattern of $Ni_{21}Sn_2P_6$. The lower grey plot shows the difference between the measured and refined powder pattern (Rwp = 3.55).

Table 6.1: Crystallographic data of $Ni_{21}Sn_2P_6$

Crystal Size	0.05 x 0.05 x 0.05 mm	μ	32.17 mm^{-1}
Crystal System	Cubic	observed refl.	8252
Structure Type	C_6Cr_{23}	unique refl.	279
Spacegroup	$Fm\overline{3}m$	unique refl. > 4s (Fo)	258
Density (X-ray)	7.996 g/cm^3	Refined parameters	14
Formula Units	4	R1	0.0159
Lattice Parameter	a=1112.2(2) pm	wR2	0.0352
Volume	1.37578 nm^3	remaining electron density	0.79 Å^{-3}
F(000)	3112		

Crystal structure information of this phase is given in Table 6.1. Five crystallographically different atomic positions corresponding to the arrangement in the C_6Cr_{23} type structure were identified: three Ni, one Sn and one P atom, the latter occupying the C position in C_6Cr_{23}; atomic positions are given in Table 6.2a together with anisotropic displacement parameters. Selected interatomic distances are compiled in Table 6.3. A general image of the $Ni_{21}Sn_2P_6$ crystal structure is given in Fig. 6.2.

Table 6.2a: Atomic positions and anisotropic displacement parameters of $Ni_{21}Sn_2P_6$

Atomic Positions	Site	x	y	z	U_{eq}
Ni1	48h	0	0.17442(2)	0.17442(2)	0.01492(10)
Ni2	32f	0.38356(2)	0.38356(2)	0.38356(2)	0.00645(9)
Ni3	4a	0	0	0	0.0124(2)
P	24e	0.26170(8)	0	0	0.00597(14)
Sn	8c	0.25	0.25	0.25	0.00814(9)

Atomic Positions	Anisotropic Displacement Parameters					
	U11	U22	U33	U23	U13	U12
Ni1	0.0306(2)	0.00708(10)	0.00708(10)	0.00129(9)	0	0
Ni2	0.00645(9)	0.00645(9)	0.00645(9)	0.00060(6)	0.00060(6)	0.00060(6)
Ni3	0.0124(2)	0.0124(2)	0.0124(2)	0	0	0
P	0.0072(3)	0.0054(2)	0.0054(2)	0	0	0
Sn	0.00814(9)	0.00814(9)	0.00814(9)	0	0	0

Table 6.2b: Crystallographic data of $Ni_{10}P_3Sn$ according to Ref. [82]

Space group	$P\bar{3}m1$			
Lattice Parameters	a=767.4(1) pm $\quad c$=962.1(1) pm			

Atomic Positions	Site	x	y	z
Ni1*	2d	1/3	2/3	0.3847(3)
Ni2*	2d	1/3	2/3	0.9026(3)
Ni3*	2c	0	0	0.3805(3)
Ni4*	6h	0.3382(2)	0	1/2
Ni5*	6i	0.1886(1)	-0.1886(1)	0.0719(2)
Ni6*	6i	0.1383(1)	-0.1383(1)	0.8013(2)
Ni7*	6i	0.4912(1)	-0.4912(1)	0.2407(1)
P1*	3e	1/2	0	0
P2*	6i	0.1647(2)	-0.1647(2)	0.3087(3)
Sn1*	1a	0	0	0
Sn2*	2d	1/3	2/3	0.6391(2)

The atomic arrangement in $Ni_{21}Sn_2P_6$ (C_6Cr_{23} type structure) can be described based on a network of CN (=coordination number) 16 Friauf polyhedra built from 12 Ni1 atoms (48h) and four Ni2 atoms (32f) and centered by Sn (8c). These polyhedra (shown in Fig. 6.2) are linked via common edges. The Ni3 atoms (4a) are coordinated by 12 Ni1 atoms forming a cuboctahedron, which shares common faces with the Friauf polyhedra. Finally, P atoms (on the non-metal site 24e in this structure type) are surrounded by four Ni1 and four Ni2 atoms forming a square antiprism. The network of Friauf polyhedra also forms one empty interstitial site at (½, ½, ½) surrounded by a cube of Ni2 atoms.

Table 6.3: Selected interatomic distances in $Ni_{21}Sn_2P_6$

Atoms		Distance (pm)	Atoms		Distance (pm)
Ni1 -	P x2	216.92(5)	Ni3 -	Ni1 x12	274.35(5)
	Ni1 x1	237.74(8)			
	Ni2 x4	273.92(3)	Sn -	Ni2 x4	257.29(5)
	Ni1 x4	274.35(4)		Ni1 x12	302.39(3)
	Ni3 x1	274.35(5)			
	Sn x2	302.39(3)	P -	Ni1 x4	216.92(5)
				Ni2 x4	227.84(6)
Ni2 -	P x3	227.84(6)			
	Sn x1	257.29(5)			
	Ni2 x3	259.00(5)			
	Ni1 x6	273.92(3)			

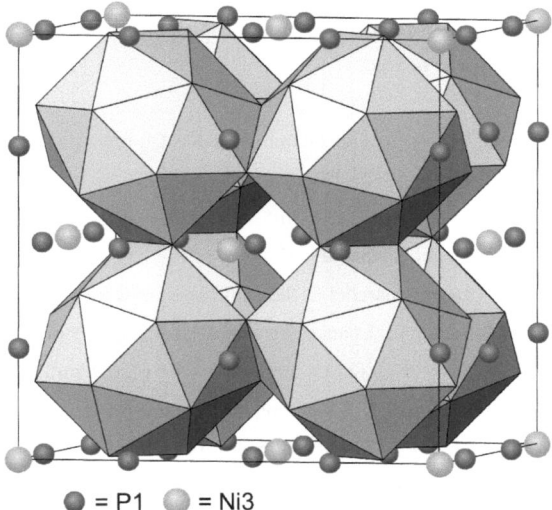

● = P1   ○ = Ni3

Fig. 6.2: General view of the $Ni_{21}P_6Sn_2$ crystal structure showing the coordination of Sn built from Ni1 and Ni2 atoms. The Ni3 atoms are shown light grey and P dark grey.

The Friauf polyhedra around Sn are thus exclusively built from Ni atoms at distances of 257.3 pm (Ni2) and 302.4 pm (Ni1). Highly anisotropic displacement parameters were observed for the Ni1 atoms, which can be explained from the rather short Ni1-Ni1 distances (237.74 pm). These are only slightly larger than twice the covalence radius of Ni (see Fig. 6.3).

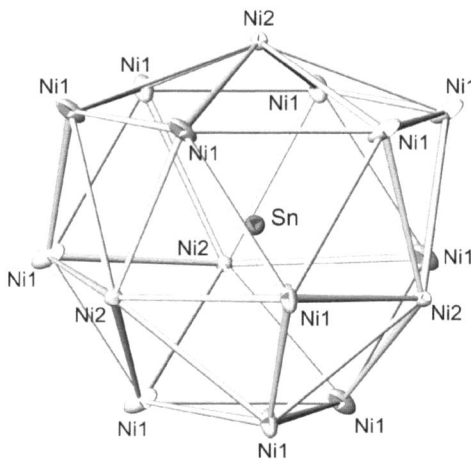

Fig. 6.3: CN 16 Friauf Polyhedron built from Ni1 and Ni2 atoms around Sn. The highly anisotropic displacement ellipsoids of Ni1 can be seen.

The cuboctahedron around Ni3 is the same type as in the Cu-type structure of pure Ni. In the ternary compound the Ni3 - Ni1 and Ni1- Ni1 distances (both 274.35 pm) found in the polyhedron are larger than the distances in pure Ni (249.6 pm), whereas the shortest distance between Ni1-Ni1 (237.74 pm) from two different cuboctahedra is shorter.

The Friauf polyhedra form channels parallel to the cubic axes, in which the Ni3 and P atoms atoms are located. However, the interatomic distances between Ni3 – P (291.5 pm) and P – P (530.1 pm) are long compared to the Ni1 – Ni3 (274.4 pm), Ni1 – P (216.9 pm) and Ni2 – P distances (227.8 pm) and thus do not represent direct contacts. The P-atoms are therefore exclusively coordinated by atoms from the CN 16 Friauf polyhedra, i.e. four Ni1 and four Ni2 atoms forming the quadratic antiprism mentioned above. These antiprisms are linked via common edges or corners, while they share common faces (i.e. one of their squares) with the cuboctahedra (made from Ni1 around Ni3).

6.3 Relation to other C_6Cr_{23} compounds

A comparison of selected ternary Ni-phosphides and -borides from the literature is given in Table 6.4 together with the lattice parameter, the variant type as mentioned in Chapter 6.1 and selected interatomic distances (only for variant 1). It can be seen that regardless of the variant type the lattice parameters are in a comparable range for the phosphides and borides, respectively, with the unit cell length of the phosphides being approx. 50 pm larger.

Table 6.4: Ternary ordered C_6Cr_{23} type compounds and some interatomic distances

Phase	Lattice Parameter [pm]	Variant	Interatomic Distances [pm]		
			$48h - 48h$	$32f - 8c$	$48h - 4a$
$Ni_{21}Sn_2P_6$	1111.2	1	237.74	257.29	274.35
$Ni_{21}Sn_2B_6$	1059(8)	1	240	239	255
$Ni_{21}In_2P_6$	1111.20(4)	1	238.9	258.3	273.42
$Mg_3Ni_{20}P_6$	1111.3(3)	2			
$Mn_3Ni_{20}P_6$	1108.5(1)	2			
$Mg_3Ni_{20}B_6$	1056.9	2			

Atomic positions:
Variant 1:   Ni: $32f$, $48h$, $4a$ P, B: $24e$ add. Element: $8c$
Variant 2:   Ni: $32f$, $48h$ P, B: $24e$ add. Element: $4a$, $8c$

Compounds of the ternary variant 1 of the C_6Cr_{23} type structure ($Ni_{21}M''_2X_6$) have the following common features:

- X atoms are exclusively coordinated by Ni atoms (quadratic antiprism)
- M'' atoms are located in CN 16 Friauf polyhedra
- A dense arrangement of Ni atoms containing fcc-fragments (cuboctahedral coordination) separates the M'' and X atoms from each other

However, even in the Friauf polyhedra, which offer sufficient space for the inclusion of large atoms, remarkably short distances are evident for the Ni2-Sn or Ni2-In bonds (considering the larger size of the Sn- or In-atoms); their length is about the same as the Ni2-Ni2 distances, in $Ni_{21}Sn_2P_6$ and $Ni_{21}Sn_2B_6$ even shorter. This observation is in agreement with the reports by Stadelmayer et al. [107], who interpret this phenomenon as a contraction of the Sn (or In) atom due to positive charging.

6.4 Comparison of the crystal structures of $Ni_{10}P_3Sn$ and $Ni_{21}Sn_2P_6$

The choice of the unit cell for a given crystal structure is in the first place arbitrary, as long as the requirement of complete space filling by simple repetition is fulfilled. Such an approach, however, is impractical, as it will in most cases lead to an unnecessarily complicated description and complicates the comparison and classification of crystal structures. Therefore a set of guidelines for the setting of crystal structures has been developed, see e.g. Ref. [110]. Nevertheless it can be desirable to describe a crystal structure by not using the simplest or conventional cell, e.g. for comparison of crystal structures. In order to be able to set up a crystal structure in a different system, it is necessary to find the geometrical relation between the axes of the original and the new unit cell in order to calculate the new cell dimensions.

As the compounds $Ni_{10}P_3Sn$ and $Ni_{21}Sn_2P_6$ have quite a small difference in composition, structural relationships between the two structures are likely. Indeed, the axes of the trigonal unit cell of $Ni_{10}P_3Sn$ can easily be related to the face and body diagonal, respectively, of the cubic $Ni_{21}Sn_2P_6$ cell:

$$a_{trig.} \approx \frac{a_{cub.}}{2} \cdot \sqrt{2} \qquad c_{trig.} \approx \frac{a_{cub.}}{2} \cdot \sqrt{3}$$

For example, the face diagonals [101] and [01-1], and the body diagonal, [-111] satisfy this requirement.

Applying this relation, the cell parameters 786 and 963 pm are obtained for a theoretical trigonal setting of $Ni_{21}Sn_2P_6$ compared to the parameters 759 and 980 pm, respectively, reported for $Ni_{10}P_3Sn$ [82].

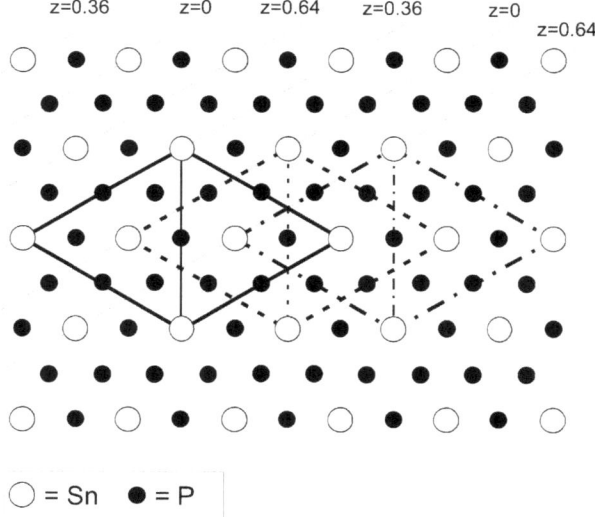

Fig. 6.4: Arrangement of the Sn and P atoms in $Ni_{10}P_3Sn$ showing the A-B-C stacking sequence of P-Sn layers (view along the *c*-axis). The individual layers are distinguished by the z-coordinates of the Sn atoms. Note that the P atoms are slightly shifted out of the layers. The building unit is given for each layer. A similar atomic arrangement can also be found in $Ni_{21}Sn_2P_6$ when viewed along the [111] direction (body diagonal). A: solid line; B: dashed line; C: dash-dotted line

Form Fig. 6.4 it can be seen that the arrangement of the P and Sn atoms is similar (but not identical) for both structures. It consists of hexagonal P – Sn nets with A-B-C stacking sequence. In both structures these nets are not flat, because the Sn atoms are situated directly within the layers and the P atoms are slightly shifted out of the plains. While in $Ni_{21}Sn_2P_6$ the distance between different layers is always one sixth of the body diagonal (corresponding to an "ideal" sequence of $z=0$, 1/3 and 2/3 in a theoretical trigonal setting), in $Ni_{10}P_3Sn$ the layer spacing is unequal ($z=0$, 0.36 and 0.64, compare the z-coordinates of $Sn1^*$ and $Sn2^*$ in that structure). Unfortunately, such a strong relation was not found for the Ni atoms, where the difference of the arrangement is reflected in the different Ni-polyhedra networks of the two compounds: a CN 16 Friauf polyhedron in $Ni_{21}Sn_2P_6$ compared to 12- and 14-fold Frank-Kasper polyhedra in $Ni_{10}P_3Sn$ [82].

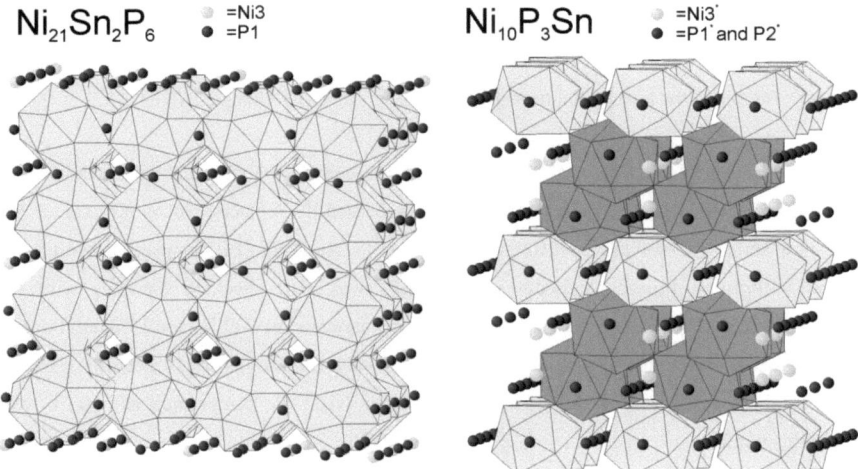

Fig. 6.5: Comparison of $Ni_{21}Sn_2P_6$ (above) and $Ni_{10}P_3Sn$ (below, view inclined along *b*-axis); Sn centred polyhedra and channels in between are shown. The different types of polyhedra in $Ni_{10}P_3Sn$ are shown in different grey shades.

The composition difference between the two phases can be seen in the atomic arrangement of the crystal structures. In $Ni_{10}P_3Sn$ (see Table 6.2b for crystal structure information according to Ref. [82]) $P1^*$ and $P2^*$ (the superscript '*' will be used to denote atom positions in $Ni_{10}P_3Sn$) are coordinated by 8 and 9 Ni atoms, respectively, in which Ni atoms from all sites contribute to build up the polyhedra. On the contrary in $Ni_{21}Sn_2P_6$ the P atoms are only coordinated by Ni1 and Ni2 (square anti-prism). The Ni3 atoms are not part of these polyhedra, but the number of Ni3 atoms in the unit cell (i.e., 4) exactly matches the difference in the stoichiometry of $Ni_{21}Sn_2P_6$ and $Ni_{10}P_3Sn$.

A similar analysis can be done on the basis of the Ni-Sn polyhedra network. When expanding the 12-fold coordination of Sn1* in Ni$_{10}$P$_3$Sn to a distorted 14-fold coordination figure by inclusion of Ni3* in the polyhedron, the Ni-Sn polyhedra networks in both Ni$_{21}$Sn$_2$P$_6$ and Ni$_{10}$P$_3$Sn have the same stoichiometry. Again in Ni$_{21}$Sn$_2$P$_6$ Ni3 is not part of the polyhedra network. Ni3 is therefore neither part of the Ni-P nor the Ni-Sn network, which is fully consistent with its exclusive coordination by other Ni atoms.

The description of the two crystal structures on the basis of the Ni-Sn network also reveals another relation between these phases. In Ni$_{21}$Sn$_2$P$_6$ Ni3 and P are located in channels formed by the CN 16 Friauf polyhedra (cf. Fig. 65). Channels formed by the Ni coordination (including Ni3*) of Sn1* and Sn2* can also be found in Ni$_{10}$P$_3$Sn, in which P1* and P2* are located in the channels. While these channels are parallel to the cubic a-axes in Ni$_{21}$Sn$_2$P$_6$, they are parallel to the trigonal a-axes in Ni$_{10}$P$_3$Sn. In both structures, however, the P – P distances in the channels are longer than the distances to Ni atoms in the polyhedral network.

According to Keimes et al. [82] in Ni$_{10}$P$_3$Sn large atoms can only be placed within the Frank-Kasper polyhedra functioning as cages. The authors successfully replaced Sn by Zn, Ga or Sb, resulting in a series of isotypic structures in addition to Ni$_{10}$P$_3$Sn. In the present work, this substitution within the Friauf polyhedra of Ni$_{21}$Sn$_2$P$_6$ was also checked. Samples with the composition Ni$_{21}$Y$_2$P$_6$ (Y=Zn, Sb, Ga) were prepared and annealed at 700 °C for 15 days. Powder diffraction revealed that none of these three samples was single phase (as it had been the case for Ni$_{21}$Sn$_2$P$_6$), and no evidence for the existence of further Ni$_{21}$Y$_2$P$_6$ compounds at 700 °C was obtained in this work.

7. Summary

7.1 Summary (English)

Today's electronics industry faces complex production chains (e.g. BGA assemblies, SMT), a high lead density on their circuit boards and high demands in reliability. Thus the quality of the solder joints has become the central issue. IMC formation in the joint during soldering therefore needs to be precisely controlled.

The system Ni-P-Sn has high technological relevance as it provides the scientific basis for the understanding, interpretation and subsequent control for the IMC formation between common ENIG (electroless Ni(P)-Au) surfaces and Sn-based solders. Within the transition to lead-free solders this has become particularly important due to the much higher Sn-content in the solders now in use.

In this thesis the Ni-P and Ni-P-Sn systems have been investigated by X-ray diffraction, electron microscopy including electron probe microanalysis and differential thermal analysis.

While the basic outline of the binary Ni-P phase diagram was not changed too much, many details were altered based on the new results. Significant homogeneity ranges were introduced for the Ni_5P_2 and $Ni_{12}P_5$ HT-phases, and in contrast to the literature reports $Ni_{12}P_5$ HT was interpreted to be a congruently melting compound. However, in agreement with the literature, it was found that none of these two phases can be retained by quenching, so that this high temperature area had to be exclusively based on DTA data. Furthermore, Ni_5P_4 was found to be formed by the peritectic reaction $L + Ni_2P = Ni_5P_4$, which contradicts the literature, and invariant reaction temperatures in the middle part of the phase diagram were determined by DTA.

In the Ni-P-Sn system four partial isothermal sections at 200, 550, 700 and 850 °C were established based on the primary XRD and EPMA data. Invariant reactions and their reaction temperatures were deduced from the DTA data. Three (partial) isopleths (vertical sections) in the Ni-rich corner, a liquidus surface and the reaction scheme (Scheil Diagram) were drawn from the combined data.

Five ternary compounds exist in the Ni-rich part of the Ni-P-Sn system, of which four have been described in the literature. $Ni_{21}P_6Sn_2$ was found during work on the ternary phase diagram. Its crystal structure was determined by single crystal X-ray diffraction to be a C_6Cr_{23} type structure, and its composition differs from neighbouring $Ni_{10}P_3Sn$ by only one atomic per cent phosphorus. It is therefore not surprising that there exist several similarities between the two crystal structures, e.g. in the arrangement of the Sn- and P-networks and the complete separation of P and Sn by Ni-atoms. Both $Ni_{21}P_6Sn_2$ and $Ni_{10}P_3Sn$ were included as congruently melting compounds. Furthermore, $Ni_{21}P_6Sn_2$ is part of a family of isotypic ternary borides, carbides and phosphides.

In the Ni-rich part most invariant reactions were found in the temperature interval from 861 to 1010 °C, i.e. at temperatures above the highest investigated isotherm. A consistent description of the phase equilibria could be established and is represented by the Scheil Diagram. A total of 27 invariant reactions have so far been identified in this region, whereas the detailed transition from the ternary solid solution based on the Ni_3Sn_2 HT-phase to the various phases forming out of it at lower temperatures has not been elucidated in detail. Phase transitions in such NiAs-type phases are known from literature to be crystallographically complex, which is is usually reflected in the phase diagram, too.

As expected, P caused a number of serious experimental problems during sample preparation and interpretation. At P-contents higher than 40 at.% the evaporation and loss of P could not be avoided, which resulted in explosions of the employed quartz capsules during the initial production runs. Furthermore, non-equilibrium samples were frequently obtained even after prolonged annealing due to the influence of the gas phase, which plays a significant role on the (P,Sn)-rich side. A metastable reaction from the binary Ni-P appears to have a counterpart in the ternary system, too. As a result, inconsistent phase field information was obtained from samples annealed and quenched from 200 °C.

Despite all experimental difficulties a consistent version of the Ni-P-Sn phase diagram could be compiled. Information from this phase diagram will be helpful in interpreting the reactive phase formation in solder joints and for the development of new materials and techniques for the electronics industry (e.g. transient liquid phase bonding). In addition, the experimental information from this work should be used as input for CALPHAD modelling, because our experiences showed that a combination of experimental work and theoretical and semi empirical modelling will be required to solve this system.

7.2 Zusammenfassung (Deutsch)

Die Elektronikindustrie vertraut heutzutage auf komplexe Produktionsabläufe (z.B. BGA Aufbauten, SMT) und auf hohe Anschlussdichten bei gleichzeitig hohen Qualitätsansprüchen. Die Qualität von Lötverbindungen ist ein zentrales Thema, und daher ist es unerlässlich, die Bildung intermetallischer Verbindungen in einer Lötstelle zu verstehen und zu beherrschen.

Das System Ni-P-Sn hat große technologsiche Bedeutung, da es die wissenschaftliche Grundlage für das Verständnis, die Interpretation und Beherrschung der Verbindungsbildung zwischen Ni(P)/Au Oberflächen und Sn-basierten Loten liefert. Durch die Umstellung auf bleifreie Lote hat dieses System nochmals an Bedeutung gewonnen, da die verwendeten bleifreien Lote einen weit höheren Sn-Gehalt haben.

In der vorliegenden Arbeit wurden die Systeme Ni-P und Ni-P-Sn mittels Röntgenbeugung, Elektronenmikroskopie (mit Elektronenstrahlmikrosonde) und Differenzthermoanalyse untersucht. Das Ni-P Phasendiagramm erfuhr zwar keine grundlegenden Änderungen, wurde aber in Bezug auf zahlreiche Details gemäß den neuen Daten gründlich überarbeitet. Für die Verbindungen Ni_5P_2 HT und $Ni_{12}P_5$ HT wurden signifikante Phasenbreiten gefunden, und im Gegensatz zur vorhandenen Literatur wurde $Ni_{12}P_5$ HT als kongruent schmelzende Verbindung interpretiert. Dies muss jedoch allein auf Daten aus DTA-Messungen beruhen, da keine der beiden HT-Phasen durch Abschrecken bei Raumtemperatur stabilisiert werden kann, was gut mit Berichten aus der Literatur übereinstimmt. Weiters wurde eine peritektische Bildung von Ni_5P_4 gemäß $L + Ni_2P = Ni_5P_4$ gefunden, was eine Änderung gegenüber der Literatur bedeutet. Im Mittelteil des Phasendiagramms wurden außerdem die Temperaturen von invarianten Reaktionen mittels DTA bestimmt.

Im ternären System Ni-P-Sn wurden ausgehend von den XRD und EPMA Daten vier Isothermen bei 200, 550, 700 und 850 °C erstellt. Invariante Reaktionen und die Reaktionstemperaturen wurden aus den DTA Messungen abgeleitet. Drei Isoplethen (vertikale Schnitte), die Liquidus-Oberfläche und das Scheil-Diagramm für den Ni-reichen Teil wurden unter Berücksichtigung aller Ergebnisse erstellt.

Fünf ternäre Verbindungen existieren im Ni-P-Sn System, von denen vier in der Literatur beschrieben worden sind. $Ni_{21}P_6Sn_2$ wurde während der Arbeit am ternären Phasendiagramm entdeckt. Die Kristallstruktur dieser Verbindung wurde mittels Einkristallröntgenbeugung als ein weiterer Vertreter des C_6Cr_{23} Strukturtyps identifiziert. Damit gehört $Ni_{21}P_6Sn_2$ zu einer ganzen Familie von ternären Boriden, Carbiden und Phosphiden. Ihre Zusammensetzung unterscheidet sich von der benachbarten Phase $Ni_{10}P_3Sn$ nur um ein Atomprozent im P-Gehalt. Erwartungsgemäß gibt es daher mehrere Gemeinsamkeiten zwischen den beiden Strukturen, z.B. in der Anordnung der Sn-

und P-Atome, und in der kompletten Abschirmung von P und Sn durch Ni-Atome. Für beide Verbindungen, $Ni_{10}P_3Sn$ und $Ni_{21}P_6Sn_2$, wurde kongruentes Schmelzverhalten gefunden.

Der Großteil der invarianten Reaktionen im Ni-reichen Teil des Phasendiagramms liegt in einem vergleichsweise engen Temperaturbereich von 861 bis 1010 °C, d.h. bei Temperaturen oberhalb der höchsten untersuchten Isotherme (850 °C). Im Scheil Diagramm ist die gefundene konsistente Abfolge dieser Reaktionen dargestellt. In Summe wurden bis jetzt 27 invariante Reaktionen allein im Ni-reichen Teil identifiziert, wobei aber der Übergang von Ni_3Sn_2 HT in die zahlreichen Tieftemperaturphasen nicht im Detail aufgeklärt werden konnte. Aus der Literatur ist bekannt, dass Phasenübergänge in Phasen vom NiAs-Typ kristallographisch sehr komplex sind, was sich auch stets im Phasendiagramm manifestiert.

Wie erwartet, erschwerte die Anwesenheit von Phosphor die Probenherstellung und auch die Interpretation der Ergebnisse. Ab einem P-Gehalt von 40 at.% konnten Verdampfung / Sublimation und der Verlust von Phosphor aus der Probe nicht verhindert werden, was anfänglich zu Explosionen der verwendeten Quartzgefäße während der Probenherstellung führte. Außerdem wurden auch nach langem Gleichgewichtsglühen häufig Proben erhalten, die das thermodynamsiche Gleichgewicht nicht erreicht hatten, was auf das Vorhandensein einer P-haltigen Gasphase vor allem im (P,Sn)-reichen Bereich zurückzuführen ist. Auch eine metastabile Reaktion im binären Ni-P dürfte eine Entsprechung im ternären System haben, wodurch widersprüchliche Phasendiagramminformation von Proben erhalten wurde, die bei 200 °C geglüht worden waren.

Trotz dieser experimentellen Schwierigkeiten konnte eine konsistente Version des Ni-P-Sn Phasendiagramms erstellt werden. Dieses Phasendiagramm wird nützlich sein für die Interpretation der Bildung von intermetallischen Verbindungen in einer Lötstelle durch chemische Reaktion, sowie auch für die Entwicklung von neuen Materialien und Techniken für die Elektronikindustrie (z.B. Diffusionslöten). Abgesehen davon sollten die experimentellen Daten auch Ausgangspunkt für eine CALPHAD-Modellierung sein, denn die gewonnenen Erfahrungen haben gezeigt, dass eigentlich nur durch die Kombination von experimentellen Methoden und Modellrechnungen die Unklarheiten in solch einem komplexen System gelöst werden können.

8. References

1. Rahn, A. *The Basics of Soldering*. 1993; John Wiley & Sons Inc.
2. RoHS, 2003. *Directive 2002/95/EC of the European Parliament and of the Council of 27 January 2003 on the restriction of the use of certain hazardous substances in electrical and electronic equipment (RoHS)*
3. Sohn, SE. Circuits Assem. 2002;13:32.
4. Suga, T. *Roadmap for the Commercialization of Lead-Free Solder* A report of the Lead-Free Soldering Roadmap Committee, JEITA. Online: <http://tsc.jeita.or.jp/tsc/comms/7_easm/english/leadfree/data/MAP-paper.doc>.
5. Suganuma, K and Kim, K-S. JOM 2008;60(6):61.
6. Sharif, A and Chan, YC. J. Alloys Compd. 2005;393(1-2):135.
7. Choubey, A; Yu, H; Osterman, M; Pecht, M; Yun, F; Li, Y and Ming, X. J. Electron. Mater. 2008;37(8):1130.
8. Noel, H. *Basic Research on Uranium Intermetallics for the Next Generation of Nuclear Fuels*. 2008; oral presentation, Inst. of Physical Chemistry, University of Vienna.
9. Tang, PT. *Fabrication of Micro Components by Electrochemical Deposition*. PhD Thesis, Institut for Procesteknik, Technical University of Denmark, 1998.
10. Yoon, J-W; Park, J-H; Shur, C-C and Jung, S-B. Microelectron. Eng. 2007;84(11):2552.
11. Kumar, A and Chen, Z. Mater. Sci. Eng., A 2006;A423(1-2):175.
12. Sharif, A; Chan, YC; Islam, MN and Rizvi, MJ. J. Alloys Compd. 2005;388(1):75.
13. Islam, MN; Chan, YC; Sharif, A and Alam, MO. Microelectron. Reliab. 2003;43(12):2031.
14. He, M; Chen, Z and Qi, G. Acta Mater. 2004;52(7):2047.
15. David, W-Y and Chan, S. Can. Pat. Appl. 2007:14.
16. Kim, G-S and Yoon, W-G. Repub. Korean Kongkae Taeho Kongbo 2004:page not available.
17. He, M; Chen, Z; Qi, G; Wong, CC and Mhaisalkar, SG. Thin Solid Films 2004;462-463:363.
18. Yoon, J-W and Jung, S-B. J. Alloys Compd. 2004;376(1-2):105.
19. Jee, YK; Yu, J and Ko, YH. J. Mater. Res. 2007;22(10):2776.
20. Yoon, J-W; Kim, S-W and Jung, S-B. J. Alloys Compd. 2004;385(1-2):192.
21. Kao, S-T and Duh, J-G. J. Electron. Mater. 2005;34(8):1129.
22. Li, JF; Mannan, SH; Clode, MP; Chen, K; Whalley, DC; Liu, C and Hutt, DA. Acta Mater. 2007;55(2):737.

23. Kumar, A; Chen, Z; Mhaisalkar, SG; Wong, CC; Teo, PS and Kripesh, V. Thin Solid Films 2006;504:410.
24. Chun, H-S; Yoon, J-W and Jung, S-B. J. Alloys Compd. 2007;439(1-2):91.
25. Lin, Y-C; Shih, T-Y; Tien, S-K and Duh, J-G. Scripta Materialia 2007;56:49.
26. Wang, SJ and Liu, CY. Scripta Materialia 2003;49:813.
27. Kumar, A; He, M and Chen, Z. Surf. Coat. Technol. 2005;198(1-3):283.
28. Huang, ML; Loeher, T; Manessis, D; Boettcher, L; Ostmann, A and Reichl, H. J. Electron. Mater. 2006;35(1):181.
29. He, M; Chen, Z and Qi, GJ. Metall. Mater. Trans. A 2005;36A(1):65.
30. Lin, Y-C; Shih, T-Y; Tien, S-K and Duh, J-G. J. Electron. Mater. 2007;36(11):1469.
31. Sohn, YC; Yu, J; Kang, SK; Shih, DY and Lee, TY. J. Mater. Res. 2004;19(8):2428.
32. Sohn, Y-C and Yu, J. J. Mater. Res. 2005;20(8):1931.
33. Sharif, A and Chan, YC. Thin Solid Films 2006;504(1-2):431.
34. Islam, MN; Chan, YC; Sharif, A and Rizvi, MJ. J. Alloys Compd. 2005;396(1-2):217.
35. Sharif, A and Chan, YC. J. Alloys Compd. 2007;440(1-2):117.
36. Furuseth, S and Fjellvaag, H. Acta Chem. Scand. 1994;48(2):134.
37. Hwang, C-w; Suganuma, K; Kiso, M and Hashimoto, S. J. Mater. Res. 2003;18(11):2540.
38. Inorganic Chrystal Structure Database, version 1.4.1., Fachinformationszentrum Karlsruhe.
39. private communication within COST Action MP0602 2008;
40. Schmetterer, C; Flandorfer, H; Luef, C; Kodentsov, A and Ipser, H. J. Electron. Mater. 2009;38(1):10.
41. Sun, P; Andersson, C; Wei, X; Cheng, Z; Shangguan, D and Liu, J. J. Alloys Compd. 2007;437(1-2):169.
42. Schmetterer, C; Flandorfer, H; Richter, KW; Saeed, U; Kauffman, M; Roussel, P and Ipser, H. Intermetallics 2007;15(7):869.
43. Nash, P and Nash, A. Bull. Alloy Phase Diagrams 1985;6(4):350.
44. Massalski, TB; Okamoto, H; Subramanian, PR and Kacprzak, L. *Binary Alloy Phase Diagrams*. 1996; ASM International.
45. Ghosh, G. Metall. Mater. Trans. A 1999;30A(6):1481.
46. Liu, HS; Wang, J and Jin, ZP. CALPHAD: Comput. Coupling Phase Diagrams Thermochem. 2005;28(4):363.
47. Havlicek, A. Ph.D. Thesis, University of Vienna, 1991.
48. Mikulas, W and Thomassen, L. Transactions AIME, Inst. Metal. Div. : 1937;124:111.
49. Schubert, K; Burkhardt, W; Esslinger, P; Gunzel, E; Meissner, HG; Schutt, W; Wegst, J and Wilkens, M. Naturwissenschaften 1956;43:248.

References

50. Leineweber, A; Mittemeijer, EJ; Knapp, M and Baehtz, C. Mater. Sci. Forum 2004;443-444(EPDIC 8):247.
51. Leineweber, A; Ellner, M and Mittemeijer, EJ. J. Solid State Chem. 2001;159(1):191.
52. Leineweber, A; Oeckler, O and Zachwieja, U. J. Solid State Chem. 2004;177(3):936.
53. Leineweber, A. J. Solid State Chem. 2004;177(4-5):1197.
54. Fetz, E and Jette, ER. J. Chem. Phys. 1936;4:537.
55. Nial, O. Sven. Kem. Tidskr. 1947;59:172.
56. Lihl, F and Kirnbauer, H. Monatsh. Chem. 1955;86:745.
57. Bhargava, MK and Schubert, K. J. Less-Common Metals 1973;33(2):181.
58. Lee, KJ and Nash, P, eds. *Phase Diagrams of Binary Nickel Alloys*. ed. P. Nash. 1991, ASM International: Materials Park, Ohio.
59. Jolibois, P. Comptes Rendues 1910;150:106.
60. Scholder, R; Apel, A and Haken, HL. Z. Anorg. Allg. Chem. 1937;232:1.
61. Konstantinov, N. Z. Anorg. Allg. Chemie 1908;60:405.
62. Nowotny, H and Henglein, E. Zeitschrift fuer Physikalische Chemie 1938;B40:281.
63. Koeneman, J and Metcalfe, AG. Trans. Metall. Soc. AIME 1958;212:571.
64. Yupko, LM; Svirid, AA and Muchnik, SV. Poroshk. Metall. (Kiev) 1986;(9):78.
65. Oryshchyn, S; Babizhetskyy, V; Chykhriy, S; Aksel'rud, L; Stoyko, S; Bauer, J; Guerin, R and Kuz'ma, Y. Inorg. Mater. 2004;40(4):380.
66. Aronsson, B. Acta Chem. Scand. 1955;9:137.
67. Rundqvist, S; Hassler, E and Lundvik, L. Acta Chem. Scand. 1962;16:242.
68. Saini, GS; Calvert, LD and Taylor, JB. Can. J. Chem. 1964;42(7):1511.
69. Rundqvist, S and Larsson, E. Acta Chem. Scand. 1959;13:551.
70. Rundqvist, S. Acta Chem. Scand. 1962;16:992.
71. Larsson, E. Ark. Kemi 1965;23(32):335.
72. Elfstrom, M. Acta Chem. Scand. 1965;19(7):1694.
73. Biltz, W and Heimbrecht, M. Z. Anorg. Allg. Chem. 1938;237:132.
74. Rundqvist, S and Ersson, NO. Ark. Kemi 1968;30(10):103.
75. Shim, J-H; Chung, H-J and Lee, DN. J. Alloys Compd. 1999;282(1-2):175.
76. Vivian, AC. J. Inst. Met. 1920;23:325.
77. Olofsson, O. Acta Chemica Scandinavica (1947-1973) 1970;24(4):1153.
78. Eckerlin, P and Kischio, W. Z. Anorg. Allg. Chem. 1968;363(1-2):1.
79. Zaikina, JV; Kovnir, KA; Sobolev, AN; Presniakov, IA; Kytin, VG; Kulbachinskii, VA; Olenev, AV; Lebedev, OI; Van Tendeloo, G; Dikarev, EV and Shevelkov, AV. Chem. Mater. 2008;20(7):2476.

80. Gullman, J and Olofsson, O. J. Solid State Chem. 1972;5(3):441.
81. Katz, G; Kohn, JA and Broder, JD. Acta Crystallogr. 1957;10:607.
82. Keimes, V; Blume, HM and Mewis, A. Z. Anorg. Allg. Chem. 1999;625(2):207.
83. Garcia-Garcia, FJ; Larsson, AK and Furuseth, S. J. Solid State Chem. 2002;166(2):352.
84. Garcia-Garcia, FJ; Larsson, AK and Furuseth, S. Solid State Sci. 2003;5(1):205.
85. Furuseth, S and Fjellvaag, H. Acta Chem. Scand., Ser. A 1985;A39(8):537.
86. Furuseth, S; Larsson, AK and Withers, RL. J. Solid State Chem. 1998;136(1):125.
87. Villars P. and D., CL. *Pearson's Handbook of Crystallographic Data For Intermetallic Phases*. 1991; ASM International.
88. Sheldrick, GM. SHELX-97, a program for crystal structure refinement. University of Göttingen, Germany, 1997.
89. Schmetterer, C; Vizdal, J and Ipser, H. Intermetallics 2009;accepted for publication
90. Schmetterer, C; Vizdal, J; Kroupa, A; Kodentsov, A and Ipser, H. J. Electron. Mater. 2009; submitted for publication
91. Schmetterer, C and Ipser, H. 2009; ready for submission
92. Snugovsky, L; Perovic, DD and Rutter, JW. Mater. Sci. Technol. 2000;16(9):968.
93. Dityatyev, OA; Smidt, P; Stefanovich, SY; Lightfoot, P; Dolgikh, VA and Opperman, H. Solid State Sci. 2004;6(9):915.
94. Parasyuk, OV; Olekseyuk, ID; Morenko, AO and Gorgut, GP. Pol. J. Chem. 1999;73(5):765.
95. Schmetterer, C and Ipser, H. work in progress 2009;
96. Schmetterer, C; Flandorfer, H; Richter, KW and Ipser, H. J. Electron. Mater. 2007;36(11):1415.
97. Schmetterer, C; Flandorfer, H and Ipser, H. Acta Mater. 2007;56(2):155.
98. Ipser, H; Semenova, O and Krachler, R. J. Alloys Compd. 2002;338(1-2):20.
99. Huang, Y; Yuan, W; Qiao, Z; Semenova, O; Bester, G and Ipser, H. J. Alloys Compd. 2008;458(1-2):277.
100. Schmetterer, C; Wildner, M; Giester, G; Richter, KW and Ipser, H. Z. Anorg. Allg. Chem. 2009;635(2):301.
101. Westgren, A. Jernkontorets Ann. 1933;117:501.
102. Meinhardt, D and Krisement, O. Arch. Eisenhuettenwes. 1962;33:493.
103. Yakel, HL. Acta Crystallogr., Sect. B: Struct. Sci. 1987;B43(3):230.
104. Westgren, A. Nature (London, U. K.) 1933;132:480.
105. Khan, Y and Wibbeke, H. Z. Metallkd. 1991;82(9):703.
106. Idzikowski, B and Szajek, A. J. Optoelectron. Adv. Mater. 2003;5(1):239.

107. Stadelmaier, HH; Draughn, RA and Hofer, G. Z. Metallkd. 1963;54:640.
108. Keimes, V and Mewis, A. Z. Anorg. Allg. Chem. 1992;618:35.
109. Andersson-Soederberg, M and Andersson, Y. J. Solid State Chem. 1990;85(2):315.
110. Hahn (ed.), T. *International Tables for Crystallography, Volume A*. 1996, Dordrecht, Boston, London; Kluwer Academic Publishers.

9. Appendices

9.1 List of Figures:

Fig. 1.1: Selected properties and parameters influencing the soldering process 2

Fig. 1.2: Image of a solder tip of a commercial 20W soldering iron destroyed by use with Sn 3.8Ag 0.7Cu solder. A comparison with the length of a new tip shows the amount of material dissolved during soldering. There was no mechanical influence in the destruction of the tip. 3

Fig. 1.3: Solder ball attachment on a BGA substrate (from Ref. [6]). 4

Fig. 1.4: Schematic showing the influence of the phase diagram on the whole soldering process from alloy selection via testing of the method to the end of life treatment 6

Fig. 1.5: Back scattered SEM image of Sn-3.5Ag/Ni-P/Cu interface directly after reflow soldering at 250 °C for 60s [23] 8

Fig. 1.6: Line scanned SEM image of a Cu/electroless Ni-P/Sn-3.5Ag interface after aging at 200 °C for 48h [11]. 8

Fig. 2.1: Binary Ni-Sn phase diagram according to Ref. [42]. 11

Fig. 2.2: Section of the Ni-P-Sn phase diagram at (a) 1125 K, (b) 975 K and (c) 295 K (judged from slowly cooled samples). [36] 16

Fig. 3.1: Effect of P-evaporation during annealing: quenching resulted in the condensation of a huge amount of P on the inner quartz glass wall (Ni-P alloy containing 70 at.% P). 18

Fig. 3.2: Ni-P sample annealed at 900 °C. The sample is hollow due to evaporation of P. 19

Fig. 3.3: Bruker D8 powder diffractometer used for phase analysis. The X-ray tube is on the left. The autosampler is in the center and the detector on the right. During the measurement the autosampler rotates by the angle θ, while the detector simultaneously rotates by 2θ (Bragg-Brentano geometry). 20

Fig. 3.4: Drawing and image of the experimental DTA setup used in the present study. 22

Fig. 3.5: Destroyed transducer unit of the DTA instrument after explosion of a Ni-P sample. 23

Fig. 4.1: Ni-P Phase Diagram according to the present study; the P-rich part is only tentative. 33

Fig. 4.2: Ni-rich section of the Ni-P Phase Diagram with data points from DTA: x, invariant effects; Δ, other effects. 33

Fig. 4.3: SEM image of sample NP 13 ($Ni_{68}P_{32}$) cooled down from the melt on air. The microstructure resulting from the eutectic e11, L = $Ni_{12}P_5$ HT + Ni_2P can be seen. Different shades in the light and dark phases are due to different orientation of the individual grains. 35

Fig. 4.4: Central part of the Ni-P Phase Diagram with data points from DTA as in Fig. 4.2. 37

Fig. 4.5: SEM image of sample NP 21 ($Ni_{58}P_{42}$) quenched from 900 °C showing the primary crystallization of Ni_2P, the peritectic rim of Ni_5P_4 and the eutectic matrix (Ni_5P_4 + NiP according to XRD). 37

Fig. 4.6: Section of the DTA heating curve of sample NP 15 ($Ni_{55}P_{45}$) using a heating rate of 0.1 K/min: three thermal effects can be distinguished as explained in the text. 39

Fig. 4.7: Ni-P samples NP 33, 34 and 38 with 67, 69 and 74 at.% P (from left to right); these samples were prepared from powders of Ni and red P, pressed into pellets and annealed at 700 °C: their clearly non-metallic appearance is evident. 40

Fig. 5.1a: Partial isothermal section of the Ni-P-Sn system at 850 °C. '+' indicate nominal compositions of samples; for clarity not all sample positions are indicated in this Fig. (sample positions around T1 and T2 are shown in Fig. 5.1b). Uncertain phase equilibria and L-apexes of three-phase fields are shown by dashed lines. The grey dotted rectangle indicates the area shown in Fig. 5.1b. 65

Fig. 5.1b: Enlarged area around the compounds T1 and T2 of the isotherm at 850 °C. Symbols as in Fig. 5.1a. 66

Fig. 5.2: Partial isothermal section of the Ni-P-Sn system at 700 °C. Symbols as in Fig. 5.1. 68

Fig. 5.3a: Isothermal section of the system Ni-P-Sn at 550 °C. Ambiguous phase equilibria or those added without experimental basis are shown using dashed lines. Symbols as in Fig. 5.1. An enlarged image of the Ni-rich part is shown in Fig. 5.3b. 69

Fig. 5.3b: Enlarged image of the Ni-rich part of the isotherm at 550 °C. Symbols as in Fig. 5.1. 69

Fig. 5.4: SEM images of sample NPS 29 ($Ni_{56.25}P_{25}Sn_{18.75}$) annealed at 850 (left) and 700 °C (right). Different microstructures composed of different phases according to the three phase fields [L + Ni_3Sn_2 HT + Ni_2P] (850 °C) and [Ni_2P + T3 + T5] (700 °C) were obtained. On quenching from 850 °C the liquid decomposed into several phases, (Sn), Ni_3Sn_4 and T5. The amount of this latter phase was usually small and was in this sample only found by XRD.. 70

Fig. 5.5: Scheil Diagram of the Ni-rich area of the Ni-P-Sn phase diagram.   T1 = $Ni_{10}P_3Sn$, T2 = $Ni_{21}P_6Sn_2$ 76

Fig. 5.6: a) Liquidus projection of the Ni-rich part of the system Ni-P-Sn showing binary and ternary invariant reactions and their connections via liquidus valleys. The primary crystallization is indicated by the framed text. b) Enlarged part around the primary crystallization fields of T1 and T2. 82

Fig. 5.7: SEM image of as-cast sample NPS 28. The primary crystallization of Ni_3Sn_2 HT and the eutectic matrix of Ni_3Sn_2 HT + Ni_2P can be seen. 83

Fig. 5.8: SEM image of as-cast sample NPS 24. The primary crystallization of Ni_3Sn_2 HT and the eutectic matrix of Ni_3Sn_2 HT + $Ni_{12}P_5$ LT can be seen. 84

Fig.5.9: Isopleth from Ni_3Sn to Ni_2P with data points from thermal analysis. x: invariant effect, open triangle: monovariant effect, full triangle: liquidus. Phase field designations: 85

Fig. 5.10: Detail of the isopleth from Ni_3Sn to Ni_2P between 15 and 28 at.% P. The development of the phase equilibria around the close lying reactions U12, E4 and U7 as well as e15 and e16 related to the $Ni_{12}P_5$ HT – LT transition are shown in this Figure. Symbols as in Fig. 5.11. Phase field designations: 86

Fig. 5.11: Partial Isopleth from Ni_3Sn to P until 40 at.% P. Symbols as in Fig. 5.1. Phase field designations: 87

Fig. 5.12: Isopleth from Ni_3Sn_2 HT to Ni_2P. Symbols as in Fig. 5.11. Phase field designations: 88

Fig. 5.13: SEM image of as cast sample NPS 56. The black grains of Ni_2P are considered the primary crystallization, wheras the dark grey grains are the secondary crystallization of $Ni_{12}P_5$ HT (LT in XRD). The fine matrix was interpreted to contain the three phases Ni_3Sn_2 HT + $Ni_{12}P_5$ LT + Ni_2P. 89

Fig. 5.14: DTA recording of sample P_2Sn_{98} carried out at 2K/min with Sn as reference. The sequence of peaks in the heating curve (green) indicates a eutectic reaction. 95

Fig. 5.15: SEM images of samples a) NPS 72 and b) NPS 77 quenched after annealing at 550 °C. In a) large grains of P_3Sn_4 and smaller ones of NiP_3 as well as the matrix from (Sn) and P_3Sn_4 can be seen. b) shows NiP_2 and the matrix from (Sn) and P_3Sn_4 (the black areas are holes). While the large grains indicate that P_3Sn_4 is an equilibrium phase at the annealing temperature, P_3Sn_4 in the matrix is a product of decomposition of the liquid during quenching. 97

Fig. 5.16: Partial isothermal section at 200 °C. The part that was adapted from the phase equilibria at 550 °C is shown in grey. 98

Fig. 5.17: SEM images of samples a) NPS 66 and b) NPS 75 quenched from 200 °C. Non-equilibrium effects due to incomplete reaction L + Ni$_2$P = T5 can be seen in both micrographs. Furthermore sample 75 contains four phases that do not correspond to the phase triangulation, either. 100

Fig. 6.1: Powder diffraction pattern of Ni$_{21}$Sn$_2$P$_6$. The lower grey plot shows the difference between the measured and refined powder pattern (Rwp = 3.55). 103

Fig. 6.2: General view of the Ni$_{21}$P$_6$Sn$_2$ crystal structure showing the coordination of Sn built from Ni1 and Ni2 atoms. The Ni3 atoms are shown light grey and P dark grey. 105

Fig. 6.3: CN 16 Friauf Polyhedron built from Ni1 and Ni2 atoms around Sn. The highly anisotropic displacement ellipsoids of Ni1 can be seen. 106

Fig. 6.4: Arrangement of the Sn and P atoms in Ni$_{10}$P$_3$Sn showing the A-B-C stacking sequence of P-Sn layers (view along the *c*-axis). The individual layers are distinguished by the z-coordinates of the Sn atoms. Note that the P atoms are slightly shifted out of the layers. The building unit is given for each layer. A similar atomic arrangement can also be found in Ni$_{21}$Sn$_2$P$_6$ when viewed along the [111] direction (body diagonal). A: solid line; B: dashed line; C: dash-dotted line 108

Fig. 6.5: Comparison of Ni$_{21}$Sn$_2$P$_6$ (above) and Ni$_{10}$P$_3$Sn (below, view inclined along *b*-axis); Sn centred polyhedra and channels in between are shown. The different types of polyhedra in Ni$_{10}$P$_3$Sn are shown in different grey shades. 109

9.2 List of Tables

Table 2.1: Solid phases in the binary Ni-Sn system according to Ref. [42] 12

Table 2.2: Invariant Reactions in the system Ni-Sn according to the literature [42] 13

Table 4.1: Experimental results of the phase analysis in the system Ni-P 24

Table 4.2: Experimental results of the thermal analysis in the system Ni-P 29

Table 4.3: Maximum stability ranges of binary Ni-P phases 31

Table 4.4: Invariant Reactions in the System Ni-P according to the present work and the literature. 32

Table 5.1: Experimental results of the phase analysis in the system Ni-P-Sn 41

Table 5.2: Space group, melting range and composition of ternary Ni-P-Sn Phases 61

Table 5.3: Experimental results of the thermal analysis in the system Ni-P-Sn. All samples were measured in evacuated quartz crucibles at a heating rate of 5K/min. 72

Table 5.4: Invariant reactions in the system Ni-P-Sn 79

Table 5.5: Results of the analysis of as-cast samples; all samples were air cooled from 1180 °C. '*' denotes the primary crystallization. 80

Table 5.6: Preliminary results of the phase analysis in the system P-Sn 93

Table 5.7: Experimental results of the thermal analysis in the system P-Sn. 94

Table 6.1: Crystallographic data of $Ni_{21}Sn_2P_6$ 103

Table 6.2a: Atomic positions and anisotropic displacement parameters of $Ni_{21}Sn_2P_6$ 104

Table 6.2b: Crystallographic data of $Ni_{10}P_3Sn$ according to Ref. [82] 104

Table 6.3: Selected interatomic distances in $Ni_{21}Sn_2P_6$ 105

Table 6.4: Ternary ordered C_6Cr_{23} type compounds and some interatomic distances 107

Die VDM Verlagsservicegesellschaft sucht für wissenschaftliche Verlage abgeschlossene und herausragende

Dissertationen, Habilitationen, Diplomarbeiten, Master Theses, Magisterarbeiten usw.

für die kostenlose Publikation als Fachbuch.

Sie verfügen über eine Arbeit, die hohen inhaltlichen und formalen Ansprüchen genügt, und haben Interesse an einer honorarvergüteten Publikation?

Dann senden Sie bitte erste Informationen über sich und Ihre Arbeit per Email an *info@vdm-vsg.de*.

Sie erhalten kurzfristig unser Feedback!

VDM Verlagsservicegesellschaft mbH
Dudweiler Landstr. 99   Telefon +49 681 3720 174
D - 66123 Saarbrücken   Fax   +49 681 3720 1749
www.vdm-vsg.de

Die VDM Verlagsservicegesellschaft mbH vertritt

Printed by Books on Demand GmbH, Norderstedt / Germany